Air Pollution: Causes, Impacts and Control

Edited by
Vincent King

Air Pollution: Causes, Impacts and Control
Edited by Vincent King
ISBN: 978-1-63549-021-3 (Hardback)

Published by Larsen and Keller Education,
5 Penn Plaza,
19th Floor,
New York, NY 10001, USA

Cataloging-in-Publication Data

Air pollution : causes, impacts and control / edited by Vincent King.
p. cm.
Includes bibliographical references and index.
ISBN 978-1-63549-021-3
1. Air--Pollution. 2. Air quality. I. King, Vincent.
TD883 .A47 2017
628.53--dc23

Printed and bound in the United States of America.

For more information regarding Larsen and Keller Education and its products, please visit the publisher's website www.larsen-keller.com

Table of Contents

Preface

Air pollution is a serious problem that we as a society face today. It refers to the pollution of air by emission of biological molecules, particulates and other dangerous substances into the atmosphere. The main causes of air pollution are the emissions of poisonous gases from industries, a/cs and cars, like carbon monoxide, methane, carbon dioxide, etc. This book is a compilation of chapters that discuss the most vital concepts in the field of air pollution. It includes detailed explanation of the various concepts and methods applied to control air pollution. This textbook presents the complex subject of air pollution in the most comprehensible and easy to understand language. It is an essential guide for both academicians and those who wish to pursue this discipline further.

To facilitate a deeper understanding of the contents of this book a short introduction of every chapter is written below:

Chapter 1- This chapter concentrates on providing an overview of air pollution. Air pollution causes a number of issues related to the environment and human health. Generally any substance that harms the atmosphere and causes detrimental effects to it is considered to be a pollutant of the air. The chapter on air pollution offers an insightful focus, keeping in mind the complex subject matter.

Chapter 2- A thorough account of air pollution and its causes is elucidated in this chapter. There are several causes of air pollution that have been noted, some of them being smoke, dust and volatile organic compounds. Air pollution commonly has two sources, which can be categorized as anthropogenic (man – made) sources and natural sources. This chapter elucidates the crucial theories related to air pollution.

Chapter 3- The major components of air pollution are discussed in this chapter. It deals with air pollutants, which are substances in the air that can have adverse effects on humans as well as the ecosystem. Pollutants can be natural or man-made, and are classified as primary or secondary. This chapter encompasses and incorporates the major components and key concepts of air pollution, and provides a complete understanding.

Chapter 4- The effects of air pollution can be very harsh, both on the human health as well as ecology. Air pollution is a concern equally for an individual and an entire population. It is a significant risk factor for a number of pollution-related diseases and health conditions including respiratory infections, heart disease, stroke and lung cancer.

Chapter 5- Air pollution can be best understood in confluence with the major topics listed in the following chapter. The ozone layer blankets the earth and filters harmful ultra violent radiation and is responsible for maintaining global temperature. The rays emitted back to earth causes skin and eye related problems and also has the capability to affect crops. This topic is of great importance to have a better understanding of air pollution.

Chapter 6- Atmospheric dispersion modeling is the explanation of how air pollutants are dispersed in the atmosphere. It can be used to predict future issues caused because of specific scenarios. Some of the topics explained in this text are outline of air pollution dispersion and roadway air dispersion modeling. This section has been carefully written to provide an easy understanding of air pollution analyzers.

Chapter 7- This chapter explains the measurements taken by the governments for air pollution. A continuous increase in Air Quality Index (AQI) ensures a large percentage of the population to experience severe health effects. Every country has its own air quality index and some of these are the Air Quality Health Index (Canada), the Air Pollution Index (Malaysia), and the Pollutant Standards Index (Singapore). To improve the quality of air, we certainly need to implement the measurements taken by our governments.

Chapter 8- Devices and techniques are an important component of any field of study. The following chapter elucidates the various devices that can be used to fight air pollution. The ones introduced in this chapter are cyclonic separation, electrostatic precipitator, dust collector and wet scrubber. This chapter serves as a source to understand these devices better.

Chapter 9- This chapter on air pollution offers an insightful focus, keeping in mind the complex subject matter. Some techniques to mitigate air pollution are discussed here. Geothermal heat pump and seasonal thermal energy storage are two alternative pollution alleviants that this chapter examines. With the help of both, energy can be stored and utilized in a better manner.

Chapter 10- A toxic hotspot is a location where radiation/ discharge from definite sources exposes the local population to high health risks. In urban areas, toxic hotspots can be usually found around polluted emitters such as old factories and waste storage sites. The topic discussed in this chapter are of great importance to broaden the existing knowledge on air pollution.

Finally, I would like to thank the entire team involved since the inception of this book for their valuable time and contribution. This book would not have been possible without their efforts. I would also like to thank my friends and family for their constant support.

Editor

1

Introduction to Air Pollution

This chapter concentrates on providing an overview of air pollution. Air pollution causes a number of issues related to the environment and human health. Generally any substance that harms the atmosphere and causes detrimental effects to it is considered to be a pollutant of the air. The chapter on air pollution offers an insightful focus, keeping in mind the complex subject matter.

Air pollution is the introduction of particulates, biological molecules, or other harmful materials into Earth's atmosphere, causing diseases, allergies, death to humans, damage to other living organisms such as animals and food crops, or the natural or built environment. Air pollution may come from anthropogenic or natural sources.

Air pollution from a fossil-fuel power station

The atmosphere is a complex natural gaseous system that is essential to support life on planet Earth.

Indoor air pollution and urban air quality are listed as two of the world's worst toxic pollution problems in the 2008 Blacksmith Institute World's Worst Polluted Places report. According to the 2014 WHO report, air pollution in 2012 caused the deaths of around 7 million people worldwide, an estimate roughly matched by the International Energy Agency.

Pollutants

An air pollutant is a substance in the air that can have adverse effects on humans and the ecosystem. The substance can be solid particles, liquid droplets, or gases. A pollutant can be of natural origin or man-made. Pollutants are classified as primary or secondary. Primary pollutants are usually produced from a process, such as ash from a volcanic eruption. Other examples include carbon monoxide gas from motor vehicle exhaust, or the sulfur dioxide released from factories. Secondary pollutants are not emitted directly. Rather, they form in the air when primary pollutants react or interact. Ground level ozone is a prominent example of a secondary pollutant. Some pollutants may be both primary and secondary: they are both emitted directly and formed from other primary pollutants.

Before flue-gas desulfurization was installed, the emissions from this power plant in New Mexico contained excessive amounts of sulfur dioxide.

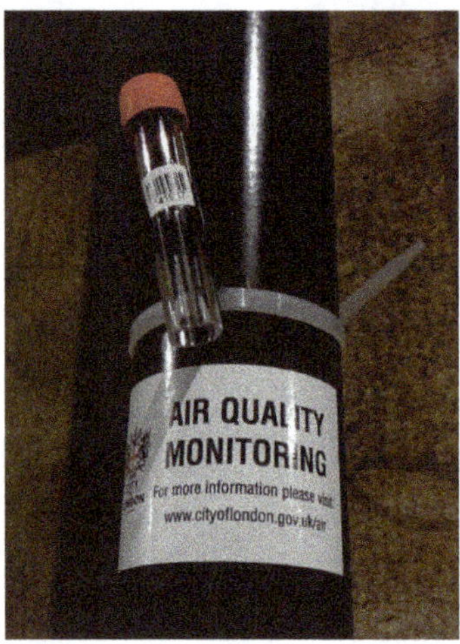

Nitrogen dioxide diffusion tube for air quality monitoring. Positioned in London City.

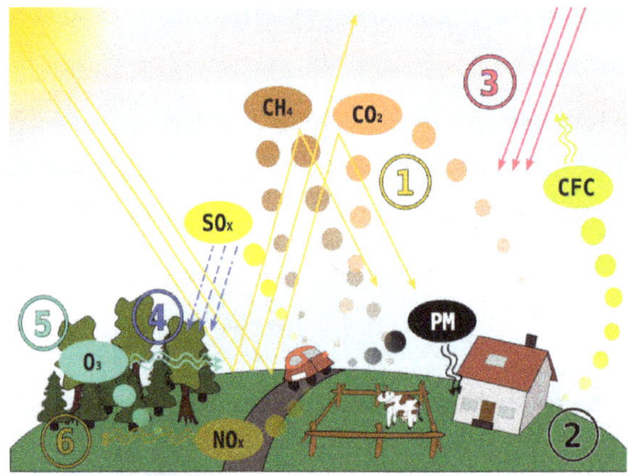

Schematic drawing, causes and effects of air pollution: (1) greenhouse effect, (2) particulate contamination, (3) increased UV radiation, (4) acid rain, (5) increased ground level ozone concentration, (6) increased levels of nitrogen oxides.

Major primary pollutants produced by human activity include:

- Sulfur oxides (SO_x) - particularly sulfur dioxide, a chemical compound with the formula SO_2. SO_2 is produced by volcanoes and in various industrial processes. Coal and petroleum often contain sulfur compounds, and their combustion generates sulfur dioxide. Further oxidation of SO_2, usually in the presence of a catalyst such as NO_2, forms H_2SO_4, and thus acid rain. This is one of the causes for concern over the environmental impact of the use of these fuels as power sources.
- Nitrogen oxides (NO_x) - Nitrogen oxides, particularly nitrogen dioxide, are expelled from high temperature combustion, and are also produced during thunderstorms by electric discharge. They can be seen as a brown haze dome above or a plume downwind of cities. Nitrogen dioxide is a chemical compound with the formula NO_2. It is one of several nitrogen oxides. One of the most prominent air pollutants, this reddish-brown toxic gas has a characteristic sharp, biting odor.
- Carbon monoxide (CO) - CO is a colorless, odorless, toxic yet non-irritating gas. It is a product by incomplete combustion of fuel such as natural gas, coal or wood. Vehicular exhaust is a major source of carbon monoxide.
- Volatile organic compounds (VOC) - VOCs are a well-known outdoor air pollutant. They are categorized as either methane (CH_4) or non-methane (NMVOCs). Methane is an extremely efficient greenhouse gas which contributes to enhanced global warming. Other hydrocarbon VOCs are also significant greenhouse gases because of their role in creating ozone and prolonging the life of methane in the atmosphere. This effect varies depending on local air quality. The aromatic NMVOCs benzene, toluene and xylene are suspected carcinogens and may lead to leukemia with prolonged exposure. 1,3-butadiene is another dangerous compound often associated with industrial use.
- Particulates, alternatively referred to as particulate matter (PM), atmospheric particulate matter, or fine particles, are tiny particles of solid or liquid suspended in a gas. In contrast, aerosol refers to combined particles and gas. Some particulates occur naturally, originating from volcanoes, dust storms, forest and grassland fires, living vegetation, and sea spray. Human activities, such as the burning of fossil fuels in vehicles, power plants and various industrial processes also generate significant amounts of aerosols. Averaged worldwide, anthropogenic aerosols—those made by human activities—currently account for approximately 10 percent of our atmosphere. Increased levels of fine particles in the air are linked to health hazards such as heart disease, altered lung function and lung cancer.
- Persistent free radicals connected to airborne fine particles are linked to cardiopulmonary disease.
- Toxic metals, such as lead and mercury, especially their compounds.
- Chlorofluorocarbons (CFCs) - harmful to the ozone layer; emitted from products are currently banned from use. These are gases which are released from air conditioners, refrigerators, aerosol sprays, etc. CFC's on being released into the air rises to stratosphere. Here they come in contact with other gases and damage the ozone layer. This allows harmful

ultraviolet rays to reach the earth's surface. This can lead to skin cancer, disease to eye and can even cause damage to plants.

- Ammonia (NH_3) - emitted from agricultural processes. Ammonia is a compound with the formula NH_3. It is normally encountered as a gas with a characteristic pungent odor. Ammonia contributes significantly to the nutritional needs of terrestrial organisms by serving as a precursor to foodstuffs and fertilizers. Ammonia, either directly or indirectly, is also a building block for the synthesis of many pharmaceuticals. Although in wide use, ammonia is both caustic and hazardous. In the atmosphere, ammonia reacts with oxides of nitrogen and sulfur to form secondary particles.
- Odours — such as from garbage, sewage, and industrial processes
- Radioactive pollutants - produced by nuclear explosions, nuclear events, war explosives, and natural processes such as the radioactive decay of radon.

Secondary pollutants include:

- Particulates created from gaseous primary pollutants and compounds in photochemical smog. Smog is a kind of air pollution. Classic smog results from large amounts of coal burning in an area caused by a mixture of smoke and sulfur dioxide. Modern smog does not usually come from coal but from vehicular and industrial emissions that are acted on in the atmosphere by ultraviolet light from the sun to form secondary pollutants that also combine with the primary emissions to form photochemical smog.
- Ground level ozone (O_3) formed from NO_x and VOCs. Ozone (O_3) is a key constituent of the troposphere. It is also an important constituent of certain regions of the stratosphere commonly known as the Ozone layer. Photochemical and chemical reactions involving it drive many of the chemical processes that occur in the atmosphere by day and by night. At abnormally high concentrations brought about by human activities (largely the combustion of fossil fuel), it is a pollutant, and a constituent of smog.
- Peroxyacetyl nitrate (PAN) - similarly formed from NO_x and VOCs.

Minor air pollutants include:

- A large number of minor hazardous air pollutants. Some of these are regulated in USA under the Clean Air Act and in Europe under the Air Framework Directive
- A variety of persistent organic pollutants, which can attach to particulates

Persistent organic pollutants (POPs) are organic compounds that are resistant to environmental degradation through chemical, biological, and photolytic processes. Because of this, they have been observed to persist in the environment, to be capable of long-range transport, bioaccumulate in human and animal tissue, biomagnify in food chains, and to have potentially significant impacts on human health and the environment.

Sources

There are various locations, activities or factors which are responsible for releasing pollutants into the atmosphere. These sources can be classified into two major categories.

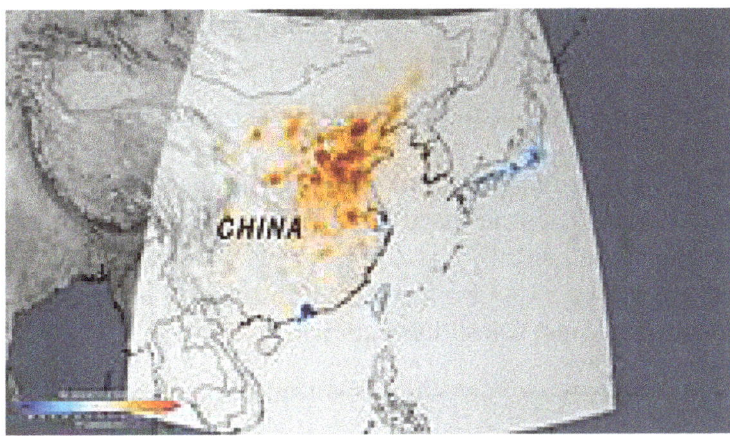

This video provides an overview of a NASA study on the human fingerprint on global air quality.

Dust storm approaching Stratford, Texas.

Controlled burning of a field outside of Statesboro, Georgia in preparation for spring planting.

Anthropogenic (Man-made) Sources:

These are mostly related to the burning of multiple types of fuel.

- Stationary sources include smoke stacks of power plants, manufacturing facilities (factories) and waste incinerators, as well as furnaces and other types of fuel-burning heating devices. In developing and poor countries, traditional biomass burning is the major source of air pollutants; traditional biomass includes wood, crop waste and dung.

- Mobile sources include motor vehicles, marine vessels, and aircraft.
- Controlled burn practices in agriculture and forest management. Controlled or prescribed burning is a technique sometimes used in forest management, farming, prairie restoration or greenhouse gas abatement. Fire is a natural part of both forest and grassland ecology and controlled fire can be a tool for foresters. Controlled burning stimulates the germination of some desirable forest trees, thus renewing the forest.
- Fumes from paint, hair spray, varnish, aerosol sprays and other solvents
- Waste deposition in landfills, which generate methane. Methane is highly flammable and may form explosive mixtures with air. Methane is also an asphyxiant and may displace oxygen in an enclosed space. Asphyxia or suffocation may result if the oxygen concentration is reduced to below 19.5% by displacement.
- Military resources, such as nuclear weapons, toxic gases, germ warfare and rocketry

Natural Sources:

- Dust from natural sources, usually large areas of land with little or no vegetation
- Methane, emitted by the digestion of food by animals, for example cattle
- Radon gas from radioactive decay within the Earth's crust. Radon is a colorless, odorless, naturally occurring, radioactive noble gas that is formed from the decay of radium. It is considered to be a health hazard. Radon gas from natural sources can accumulate in buildings, especially in confined areas such as the basement and it is the second most frequent cause of lung cancer, after cigarette smoking.
- Smoke and carbon monoxide from wildfires
- Vegetation, in some regions, emits environmentally significant amounts of Volatile organic compounds (VOCs) on warmer days. These VOCs react with primary anthropogenic pollutants—specifically, NO_x, SO_2, and anthropogenic organic carbon compounds — to produce a seasonal haze of secondary pollutants. Black gum, poplar, oak and willow are some examples of vegetation that can produce abundant VOCs. The VOC production from these species result in ozone levels up to eight times higher than the low-impact tree species.
- Volcanic activity, which produces sulfur, chlorine, and ash particulates

Emission Factors

Air pollutant emission factors are reported representative values that attempt to relate the quantity of a pollutant released to the ambient air with an activity associated with the release of that pollutant. These factors are usually expressed as the weight of pollutant divided by a unit weight, volume, distance, or duration of the activity emitting the pollutant (e.g., kilograms of particulate emitted per tonne of coal burned). Such factors facilitate estimation of emissions from various sources of air pollution. In most cases, these factors are simply averages of all available data of acceptable quality, and are generally assumed to be representative of long-term averages.

Beijing air on a 2005-day after rain (left) and a smoggy day (right)

There are 12 compounds in the list of Persistent organic pollutants. Dioxins and furans are two of them and intentionally created by combustion of organics, like open burning of plastics. These compounds are also endocrine disruptors and can mutate the human genes.

The United States Environmental Protection Agency has published a compilation of air pollutant emission factors for a multitude of industrial sources. The United Kingdom, Australia, Canada and many other countries have published similar compilations, as well as the European Environment Agency.

Air Pollution Exposure

Air pollution risk is a function of the hazard of the pollutant and the exposure to that pollutant. Air pollution exposure can be expressed for an individual, for certain groups (e.g. neighborhoods or children living in a country), or for entire populations. For example, one may want to calculate the exposure to a hazardous air pollutant for a geographic area, which includes the various microenvironments and age groups. This can be calculated as an inhalation exposure. This would account for daily exposure in various settings (e.g. different indoor micro-environments and outdoor locations). The exposure needs to include different age and other demographic groups, especially infants, children, pregnant women and other sensitive subpopulations. The exposure to an air pollutant must integrate the concentrations of the air pollutant with respect to the time spent in each setting and the respective inhalation rates for each subgroup for each specific time that the subgroup is in the setting and engaged in particular activities (playing, cooking, reading, working, etc.). For example, a small child's inhalation rate will be less than that of an adult. A child engaged in vigorous exercise will have a higher respiration rate than the same child in a sedentary activity. The daily exposure, then, needs to reflect the time spent in each micro-environmental setting and the type of activities in these settings. The air pollutant concentration in each microactivity/microenvironmental setting is summed to indicate the exposure.

Indoor Air Quality (IAQ)

A lack of ventilation indoors concentrates air pollution where people often spend the majority of their time. Radon (Rn) gas, a carcinogen, is exuded from the Earth in certain locations and trapped inside houses. Building materials including carpeting and plywood emit formaldehyde (H_2CO)

gas. Paint and solvents give off volatile organic compounds (VOCs) as they dry. Lead paint can degenerate into dust and be inhaled. Intentional air pollution is introduced with the use of air fresheners, incense, and other scented items. Controlled wood fires in stoves and fireplaces can add significant amounts of smoke particulates into the air, inside and out. Indoor pollution fatalities may be caused by using pesticides and other chemical sprays indoors without proper ventilation.

Air quality monitoring, New Delhi, India.

Carbon monoxide poisoning and fatalities are often caused by faulty vents and chimneys, or by the burning of charcoal indoors or in a confined space, such as a tent. Chronic carbon monoxide poisoning can result even from poorly-adjusted pilot lights. Traps are built into all domestic plumbing to keep sewer gas and hydrogen sulfide, out of interiors. Clothing emits tetrachloroethylene, or other dry cleaning fluids, for days after dry cleaning.

Though its use has now been banned in many countries, the extensive use of asbestos in industrial and domestic environments in the past has left a potentially very dangerous material in many localities. Asbestosis is a chronic inflammatory medical condition affecting the tissue of the lungs. It occurs after long-term, heavy exposure to asbestos from asbestos-containing materials in structures. Sufferers have severe dyspnea (shortness of breath) and are at an increased risk regarding several different types of lung cancer. As clear explanations are not always stressed in non-technical literature, care should be taken to distinguish between several forms of relevant diseases. According to the World Health Organisation (WHO), these may defined as; asbestosis, *lung cancer*, and *Peritoneal Mesothelioma* (generally a very rare form of cancer, when more widespread it is almost always associated with prolonged exposure to asbestos).

Biological sources of air pollution are also found indoors, as gases and airborne particulates. Pets produce dander, people produce dust from minute skin flakes and decomposed hair, dust mites in bedding, carpeting and furniture produce enzymes and micrometre-sized fecal droppings, inhabitants emit methane, mold forms on walls and generates mycotoxins and spores, air conditioning systems can incubate Legionnaires' disease and mold, and houseplants, soil and surrounding gardens can produce pollen, dust, and mold. Indoors, the lack of air circulation allows these airborne pollutants to accumulate more than they would otherwise occur in nature.

Health Effects

Air pollution is a significant risk factor for a number of health conditions including respiratory infections, heart disease, COPD, stroke and lung cancer. The health effects caused by air pollution may include difficulty in breathing, wheezing, coughing, asthma and worsening of existing respiratory and cardiac conditions. These effects can result in increased medication use, increased doctor or emergency room visits, more hospital admissions and premature death. The human health effects of poor air quality are far reaching, but principally affect the body's respiratory system and the cardiovascular system. Individual reactions to air pollutants depend on the type of pollutant a person is exposed to, the degree of exposure, and the individual's health status and genetics. The most common sources of air pollution include particulates, ozone, nitrogen dioxide, and sulphur dioxide. Children aged less than five years that live in developing countries are the most vulnerable population in terms of total deaths attributable to indoor and outdoor air pollution.

Mortality

The World Health Organization estimated in 2014 that every year air pollution causes the premature death of some 7 million people worldwide. India has the highest death rate due to air pollution. India also has more deaths from asthma than any other nation according to the World Health Organization. In December 2013 air pollution was estimated to kill 500,000 people in China each year. There is a positive correlation between pneumonia-related deaths and air pollution from motor vehicle emissions.

Annual premature European deaths caused by air pollution are estimated at 430,000. An important cause of these deaths is nitrogen dioxide and other nitrogen oxides (NOx) emitted by road vehicles. Across the European Union, air pollution is estimated to reduce life expectancy by almost nine months. Causes of deaths include strokes, heart disease, COPD, lung cancer, and lung infections.

The US EPA estimates that a proposed set of changes in diesel engine technology (*Tier 2*) could result in 12,000 fewer *premature mortalities*, 15,000 fewer heart attacks, 6,000 fewer emergency room visits by children with asthma, and 8,900 fewer respiratory-related hospital admissions each year in the United States.

The US EPA estimates allowing a ground-level ozone concentration of 65 parts per billion, would avert 1,700 to 5,100 premature deaths nationwide in 2020 compared with the current 75-ppb standard. The agency projects the stricter standard would also prevent an additional 26,000 cases of aggravated asthma, and more than a million cases of missed work or school.

A new economic study of the health impacts and associated costs of air pollution in the Los Angeles Basin and San Joaquin Valley of Southern California shows that more than 3,800 people die prematurely (approximately 14 years earlier than normal) each year because air pollution levels violate federal standards. The number of annual premature deaths is considerably higher than the fatalities related to auto collisions in the same area, which average fewer than 2,000 per year.

Diesel exhaust (DE) is a major contributor to combustion-derived particulate matter air pollution. In several human experimental studies, using a well-validated exposure chamber setup, DE has been linked to acute vascular dysfunction and increased thrombus formation.

The mechanisms linking air pollution to increased cardiovascular mortality are uncertain, but probably include pulmonary and systemic inflammation.

Cardiovascular Disease

A 2007 review of evidence found ambient air pollution exposure is a risk factor correlating with increased total mortality from cardiovascular events (range: 12% to 14% per 10 microg/m^3 increase).

Air pollution is also emerging as a risk factor for stroke, particularly in developing countries where pollutant levels are highest. A 2007 study found that in women, air pollution is not associated with hemorrhagic but with ischemic stroke. Air pollution was also found to be associated with increased incidence and mortality from coronary stroke in a cohort study in 2011. Associations are believed to be causal and effects may be mediated by vasoconstriction, low-grade inflammation and atherosclerosis Other mechanisms such as autonomic nervous system imbalance have also been suggested.

Lung Disease

Chronic obstructive pulmonary disease (COPD) includes diseases such as chronic bronchitis and emphysema.

Research has demonstrated increased risk of developing asthma and COPD from increased exposure to traffic-related air pollution. Additionally, air pollution has been associated with increased hospitalization and mortality from asthma and COPD.

A study conducted in 1960-1961 in the wake of the Great Smog of 1952 compared 293 London residents with 477 residents of Gloucester, Peterborough, and Norwich, three towns with low reported death rates from chronic bronchitis. All subjects were male postal truck drivers aged 40 to 59. Compared to the subjects from the outlying towns, the London subjects exhibited more severe respiratory symptoms (including cough, phlegm, and dyspnea), reduced lung function (FEV_1 and peak flow rate), and increased sputum production and purulence. The differences were more pronounced for subjects aged 50 to 59. The study controlled for age and smoking habits, so concluded that air pollution was the most likely cause of the observed differences.

It is believed that much like cystic fibrosis, by living in a more urban environment serious health hazards become more apparent. Studies have shown that in urban areas patients suffer mucus hypersecretion, lower levels of lung function, and more self-diagnosis of chronic bronchitis and emphysema.

Cancer

A review of evidence regarding whether ambient air pollution exposure is a risk factor for cancer in 2007 found solid data to conclude that long-term exposure to PM2.5 (fine particulates) increases the overall risk of non-accidental mortality by 6% per a 10 microg/m^3 increase. Exposure to PM2.5 was also associated with an increased risk of mortality from lung cancer (range: 15% to 21% per 10 microg/m^3 increase) and total cardiovascular mortality (range: 12% to 14% per a 10 microg/m^3 increase). The review further noted that living close to busy traffic appears to be associated with elevated risks of these three outcomes --- increase in lung cancer deaths, cardiovascular deaths, and overall non-accidental deaths. The reviewers also found suggestive evidence that exposure to PM2.5 is positively associated with mortality from coronary heart diseases and exposure to SO_2 increases mortality from lung cancer, but the data was insufficient to provide solid conclusions. Another investigation showed that higher activity level increases deposition fraction of aerosol particles in human lung and recommended avoiding heavy activities like running in outdoor space at polluted areas.

Cancer mainly the result of environmental factors.

In 2011, a large Danish epidemiological study found an increased risk of lung cancer for patients who lived in areas with high nitrogen oxide concentrations. In this study, the association was higher for non-smokers than smokers. An additional Danish study, also in 2011, likewise noted evidence of possible associations between air pollution and other forms of cancer, including cervical cancer and brain cancer.

In December 2015, medical scientists reported that cancer is overwhelmingly a result of environmental factors, and not largely down to bad luck. Maintaining a healthy weight, eating a healthy diet, minimizing alcohol and eliminating smoking reduces the risk of developing the disease, according to the researchers.

Children

In the United States, despite the passage of the Clean Air Act in 1970, in 2002 at least 146 million Americans were living in non-attainment areas—regions in which the concentration of certain air pollutants exceeded federal standards. These dangerous pollutants are known as the criteria pollutants, and include ozone, particulate matter, sulfur dioxide, nitrogen dioxide, carbon monoxide, and lead. Protective measures to ensure children's health are being taken in cities such as New Delhi, India where buses now use compressed natural gas to help eliminate the "pea-soup" smog. A recent study in Europe has found that exposure to ultrafine particles can increase blood pressure in children.

"Clean" Areas

Even in the areas with relatively low levels of air pollution, public health effects can be significant and costly, since a large number of people breathe in such pollutants. A 2005 scientific study for the British Columbia Lung Association showed that a small improvement in air quality (1% reduction of ambient PM2.5 and ozone concentrations) would produce $29 million in annual savings in the Metro Vancouver region in 2010. This finding is based on health valuation of lethal (death) and sub-lethal (illness) affects.

Central Nervous System

Data is accumulating that air pollution exposure also affects the central nervous system.

In a June 2014 study conducted by researchers at the University of Rochester Medical Center, published in the journal Environmental Health Perspectives, it was discovered that early exposure to air pollution causes the same damaging changes in the brain as autism and schizophrenia. The study also shows that air pollution also affected short-term memory, learning ability, and impulsivity. Lead researcher Professor Deborah Cory-Slechta said that "When we looked closely at the ventricles, we could see that the white matter that normally surrounds them hadn't fully developed. It appears that inflammation had damaged those brain cells and prevented that region of the brain from developing, and the ventricles simply expanded to fill the space. Our findings add to the growing body of evidence that air pollution may play a role in autism, as well as in other neurodevelopmental disorders." Air pollution has a more significant negative effect of males than on females.

In 2015, experimental studies reported the detection of significant episodic (situational) cognitive impairment from impurities in indoor air breathed by test subjects who were not informed about changes in the air quality. Researchers at the Harvard University and SUNY Upstate Medical University and Syracuse University measured the cognitive performance of 24 participants in three different controlled laboratory atmospheres that simulated those found in "conventional" and "green" buildings, as well as green buildings with enhanced ventilation. Performance was evaluated objectively using the widely used Strategic Management Simulation software simulation tool, which is a well-validated assessment test for executive decision-making in an unconstrained situation allowing initiative and improvisation. Significant deficits were observed in the performance scores achieved in increasing concentrations of either volatile organic compounds (VOCs) or carbon dioxide, while keeping other factors constant. The highest impurity levels reached are not uncommon in some classroom or office environments.

Agricultural Effects

In India in 2014, it was reported that air pollution by black carbon and ground level ozone had cut crop yields in the most affected areas by almost half in 2010 when compared to 1980 levels.

Historical Disasters

The world's worst short-term civilian pollution crisis was the 1984 Bhopal Disaster in India. Leaked industrial vapours from the Union Carbide factory, belonging to Union Carbide, Inc., U.S.A. (later bought by Dow Chemical Company), killed at least 3787 people and injured anywhere from 150,000 to 600,000. The United Kingdom suffered its worst air pollution event when the December 4 Great Smog of 1952 formed over London. In six days more than 4,000 died and more recent estimates put the figure at nearer 12,000. An accidental leak of anthrax spores from a biological warfare laboratory in the former USSR in 1979 near Sverdlovsk is believed to have caused at least 64 deaths. The worst single incident of air pollution to occur in the US occurred in Donora, Pennsylvania in late October, 1948, when 20 people died and over 7,000 were injured.

Alternatives to Avoid Air Pollution

There are now practical alternatives to the three principal causes of air pollution.

- Combustion of fossil fuels for space heating can be replaced by using ground source heat pumps and seasonal thermal energy storage.
- Electric power generation from burning fossil fuels can be replaced by power generation from nuclear and renewables.
- Motor vehicles driven by fossil fuels, a key factor in urban air pollution, can be replaced by electric vehicles.

Reduction Efforts

There are various air pollution control technologies and land-use planning strategies available to reduce air pollution. At its most basic level, land-use planning is likely to involve zoning and transport infrastructure planning. In most developed countries, land-use planning is an important part of social policy, ensuring that land is used efficiently for the benefit of the wider economy and population, as well as to protect the environment.

Because a large share of air pollution is caused by combustion of fossil fuels such as coal and oil, the reduction of these fuels can reduce air pollution drastically. Most effective is the switch to clean power sources such as wind power, solar power, hydro power which don't cause air pollution. Efforts to reduce pollution from mobile sources includes primary regulation (many developing countries have permissive regulations), expanding regulation to new sources (such as cruise and transport ships, farm equipment, and small gas-powered equipment such as string trimmers, chainsaws, and snowmobiles), increased fuel efficiency (such as through the use of hybrid vehicles), conversion to cleaner fuels or conversion to electric vehicles.

Titanium dioxide has been researched for its ability to reduce air pollution. Ultraviolet light will release free electrons from material, thereby creating free radicals, which break up VOCs and NOx gases. One form is superhydrophilic.

In 2014, Prof. Tony Ryan and Prof. Simon Armitage of University of Sheffield prepared a 10 meter by 20 meter-sized poster coated with microscopic, pollution-eating nanoparticles of titanium dioxide. Placed on a building, this giant poster can absorb the toxic emission from around 20 cars each day.

A very effective means to reduce air pollution is the transition to renewable energy. According to a study published in Energy and Environmental Science in 2015 the switch to 100% renewable energy in the United States would eliminate about 62,000 premature mortalities per year and about 42,000 in 2050, if no biomass were used. This would save about $600 billion in health costs a year due to reduced air pollution in 2050, or about 3.6% of the 2014 U.S. gross domestic product.

Control Devices

The following items are commonly used as pollution control devices in industry and transportation. They can either destroy contaminants or remove them from an exhaust stream before it is emitted into the atmosphere.

- **Particulate Control**
  - Mechanical collectors (dust cyclones, multicyclones)
  - Electrostatic precipitators An electrostatic precipitator (ESP), or electrostatic air cleaner is a particulate collection device that removes particles from a flowing gas (such as air), using the force of an induced electrostatic charge. Electrostatic precipitators are highly efficient filtration devices that minimally impede the flow of gases through the device, and can easily remove fine particulates such as dust and smoke from the air stream.
  - Baghouses Designed to handle heavy dust loads, a dust collector consists of a blower, dust filter, a filter-cleaning system, and a dust receptacle or dust removal system (distinguished from air cleaners which utilize disposable filters to remove the dust).
  - Particulate scrubbers Wet scrubber is a form of pollution control technology. The term describes a variety of devices that use pollutants from a furnace flue gas or from other gas streams. In a wet scrubber, the polluted gas stream is brought into contact with the scrubbing liquid, by spraying it with the liquid, by forcing it through a pool of liquid, or by some other contact method, so as to remove the pollutants.
- **Scrubbers**
  - Baffle spray scrubber
  - Cyclonic spray scrubber
  - Ejector venturi scrubber
  - Mechanically aided scrubber
  - Spray tower
  - Wet scrubber
- **Nox Control**
  - Low NOx burners
  - Selective catalytic reduction (SCR)

   - Selective non-catalytic reduction (SNCR)
   - NOx scrubbers
   - Exhaust gas recirculation
   - Catalytic converter (also for VOC control)

- **VOC Abatement**
   - Adsorption systems, using activated carbon, such as Fluidized Bed Concentrator
   - Flares
   - Thermal oxidizers
   - Catalytic converters
   - Biofilters
   - Absorption (scrubbing)
   - Cryogenic condensers
   - Vapor recovery systems

- **Acid Gas/SO_2 Control**
   - Wet scrubbers
   - Dry scrubbers
   - Flue-gas desulfurization

- **Mercury Control**
   - Sorbent Injection Technology
   - Electro-Catalytic Oxidation (ECO)
   - K-Fuel

- **Dioxin and Furan Control**
- **Miscellaneous Associated Equipment**
   - Source capturing systems
   - Continuous emissions monitoring systems (CEMS)

Regulations

In general, there are two types of air quality standards. The first class of standards (such as the U.S. National Ambient Air Quality Standards and E.U. Air Quality Directive) set maximum atmospheric concentrations for specific pollutants. Environmental agencies enact regulations which are intended to result in attainment of these target levels. The second class (such as the North American Air Quality Index) take the form of a scale with various thresholds, which is used to communicate to the public the relative risk of outdoor activity. The scale may or may not distinguish between different pollutants.

Smog in Cairo

Canada

In Canada, air pollution and associated health risks are measured with the Air Quality Health Index or (AQHI). It is a health protection tool used to make decisions to reduce short-term exposure to air pollution by adjusting activity levels during increased levels of air pollution.

The Air Quality Health Index or "AQHI" is a federal program jointly coordinated by Health Canada and Environment Canada. However, the AQHI program would not be possible without the commitment and support of the provinces, municipalities and NGOs. From air quality monitoring to health risk communication and community engagement, local partners are responsible for the vast majority of work related to AQHI implementation. The AQHI provides a number from 1 to 10+ to indicate the level of health risk associated with local air quality. Occasionally, when the amount of air pollution is abnormally high, the number may exceed 10. The AQHI provides a local air quality current value as well as a local air quality maximums forecast for today, tonight and tomorrow and provides associated health advice.

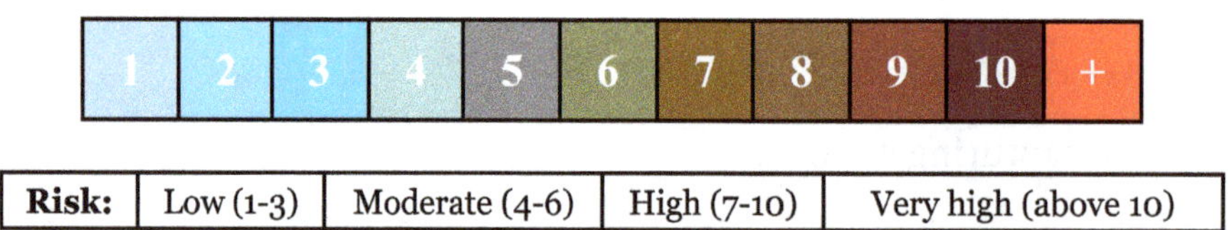

As it is now known that even low levels of air pollution can trigger discomfort for the sensitive population, the index has been developed as a continuum: The higher the number, the greater the health risk and need to take precautions. The index describes the level of health risk associated with this number as 'low', 'moderate', 'high' or 'very high', and suggests steps that can be taken to reduce exposure.

Health Risk	Air Quality Health Index	Health Messages	
		At Risk population	General Population
Low	1-3	Enjoy your usual outdoor activities.	Ideal air quality for outdoor activities
Moderate	4-6	Consider reducing or rescheduling strenuous activities outdoors if you are experiencing symptoms.	No need to modify your usual outdoor activities unless you experience symptoms such as coughing and throat irritation.
High	7-10	Reduce or reschedule strenuous activities outdoors. Children and the elderly should also take it easy.	Consider reducing or rescheduling strenuous activities outdoors if you experience symptoms such as coughing and throat irritation.
Very high	Above 10	Avoid strenuous activities outdoors. Children and the elderly should also avoid outdoor physical exertion.	Reduce or reschedule strenuous activities outdoors, especially if you experience symptoms such as coughing and throat irritation.

The measurement is based on the observed relationship of Nitrogen Dioxide (NO_2), ground-level Ozone (O_3) and particulates ($PM_{2.5}$) with mortality, from an analysis of several Canadian cities. Significantly, all three of these pollutants can pose health risks, even at low levels of exposure, especially among those with pre-existing health problems.

When developing the AQHI, Health Canada's original analysis of health effects included five major air pollutants: particulates, ozone, and nitrogen dioxide (NO2), as well as sulfur dioxide (SO_2), and carbon monoxide (CO). The latter two pollutants provided little information in predicting health effects and were removed from the AQHI formulation.

The AQHI does not measure the effects of odour, pollen, dust, heat or humidity.

Germany

TA Luft is the German air quality regulation.

Hotspots

Air pollution hotspots are areas where air pollution emissions expose individuals to increased negative health effects. They are particularly common in highly populated, urban areas, where there may be a combination of stationary sources (e.g. industrial facilities) and mobile sources (e.g. cars and trucks) of pollution. Emissions from these sources can cause respiratory disease, childhood asthma, cancer, and other health problems. Fine particulate matter such as diesel soot, which contributes to more than 3.2 million premature deaths around the world each year, is a significant problem. It is very small and can lodge itself within the lungs and enter the bloodstream. Diesel soot is concentrated in densely populated areas, and one in six people in the U.S. live near a diesel pollution hot spot.

While air pollution hotspots affect a variety of populations, some groups are more likely to be located in hotspots. Previous studies have shown disparities in exposure to pollution by race and/or income. Hazardous land uses (toxic storage and disposal facilities, manufacturing facilities, major roadways) tend to be located where property values and income levels are low. Low socioeconomic status can be a proxy for other kinds of social vulnerability, including race, a lack of ability to influence regulation and a lack of ability to move to neighborhoods with less environmental pollution. These communities bear a disproportionate burden of environmental pollution and are more likely to face health risks such as cancer or asthma.

Studies show that patterns in race and income disparities not only indicate a higher exposure to pollution but also higher risk of adverse health outcomes. Communities characterized by low socioeconomic status and racial minorities can be more vulnerable to cumulative adverse health impacts resulting from elevated exposure to pollutants than more privileged communities. Blacks and Latinos generally face more pollution than whites and Asians, and low-income communities bear a higher burden of risk than affluent ones. Racial discrepancies are particularly distinct in suburban areas of the US South and metropolitan areas of the US West. Residents in public housing, who are generally low-income and cannot move to healthier neighborhoods, are highly affected by nearby refineries and chemical plants.

Cities

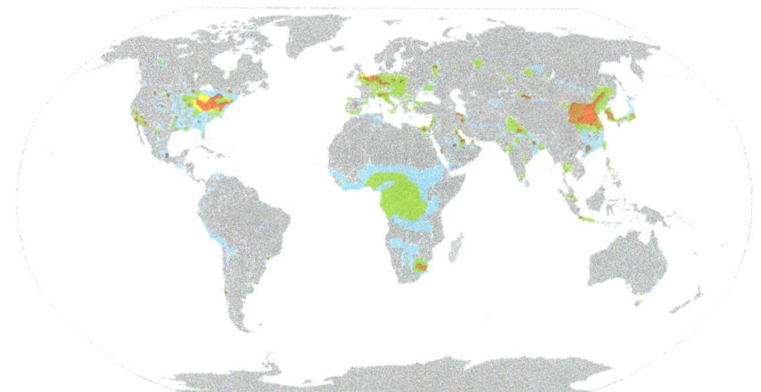

Nitrogen dioxide concentrations as measured from satellite 2002-2004

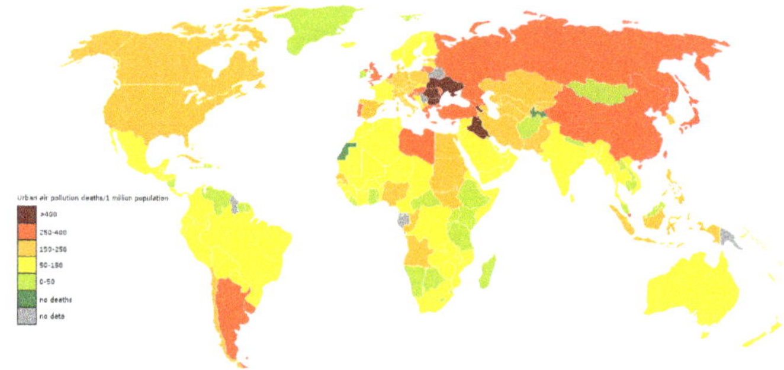

Deaths from air pollution in 2004

Air pollution is usually concentrated in densely populated metropolitan areas, especially in developing countries where environmental regulations are relatively lax or nonexistent. However, even populated areas in developed countries attain unhealthy levels of pollution, with Los Angeles and Rome being two examples. Between 2002 and 2011 the incidence of lung cancer in Beijing near doubled. While smoking remains the leading cause of lung cancer in China, the number of smokers is falling while lung cancer rates are rising.

National-scale Air Toxics Assessments 1995-2005

The national-scale air toxics assessment(NATA) is an evaluation of air toxics by the U.S. EPA. EPA has furnished four assessments that characterize nationwide chronic cancer risk estimates and noncancer hazards from inhaling air toxics. The lates was from 2005, and made publicly available in early 2011.

"EPA developed the NATA as a state-of-the-science screening tool for State/Local/Tribal Agencies to prioritize pollutants, emission sources and locations of interest for further study, in order to gain a better understanding of the risks. NATA assessments do not incorporate refined information about emission sources, but rather, use general information about sources to develop estimates of risks which are more likely to overestimate impacts than underestimate them. NATA provides estimates of the risk of cancer and other serious health effects from breathing (inhaling) air toxics in order to inform both national and more localized efforts to identify and prioritize air toxics, emission source types and locations which are of greatest potential concern in terms of contributing to population risk. This in turn helps air pollution experts focus limited analytical resources on areas and or populations where the potential for health risks are highest. Assessments include estimates of cancer and non-cancer health effects based on chronic exposure from outdoor sources, including assessments of non-cancer health effects for Diesel Particulate Matter. Assessments provide a snapshot of the outdoor air quality and the risks to human health that would result if air toxic emissions levels remained unchanged."

Most polluted cities by PM	
Particulate matter, μg/m³ (2004)	**City**
168	Cairo, Egypt
150	Delhi, India
128	Kolkata, India (Calcutta)
125	Tianjin, China
123	Chongqing, China
109	Kanpur, India
109	Lucknow, India
104	Jakarta, Indonesia
101	Shenyang, China

Governing Urban Air Pollution

In Europe, Council Directive 96/62/EC on ambient air quality assessment and management provides a common strategy against which member states can "set objectives for ambient air quality in order to avoid, prevent or reduce harmful effects on human health and the environment . . . and improve air quality where it is unsatisfactory".

On 25 July 2008 in the case Dieter Janecek v Freistaat Bayern CURIA, the European Court of Justice ruled that under this directive citizens have the right to require national authorities to implement a short term action plan that aims to maintain or achieve compliance to air quality limit values.

This important case law appears to confirm the role of the EC as centralised regulator to European nation-states as regards air pollution control. It places a supranational legal obligation on the UK to protect its citizens from dangerous levels of air pollution, furthermore superseding national interests with those of the citizen.

In 2010, the European Commission (EC) threatened the UK with legal action against the successive breaching of PM10 limit values. The UK government has identified that if fines are imposed, they could cost the nation upwards of £300 million per year.

In March 2011, the Greater London Built-up Area remains the only UK region in breach of the EC's limit values, and has been given 3 months to implement an emergency action plan aimed at meeting the EU Air Quality Directive. The City of London has dangerous levels of PM10 concentrations, estimated to cause 3000 deaths per year within the city. As well as the threat of EU fines, in 2010 it was threatened with legal action for scrapping the western congestion charge zone, which is claimed to have led to an increase in air pollution levels.

In response to these charges, Boris Johnson, Mayor of London, has criticised the current need for European cities to communicate with Europe through their nation state's central government, arguing that in future "A great city like London" should be permitted to bypass its government and deal directly with the European Commission regarding its air quality action plan.

This can be interpreted as recognition that cities can transcend the traditional national government organisational hierarchy and develop solutions to air pollution using global governance networks, for example through transnational relations. Transnational relations include but are not exclusive to national governments and intergovernmental organisations, allowing sub-national actors including cities and regions to partake in air pollution control as independent actors.

Particularly promising at present are global city partnerships. These can be built into networks, for example the C40 Cities Climate Leadership Group, of which London is a member. The C40 is a public 'non-state' network of the world's leading cities that aims to curb their greenhouse emissions. The C40 has been identified as 'governance from the middle' and is an alternative to intergovernmental policy. It has the potential to improve urban air quality as participating cities "exchange information, learn from best practices and consequently mitigate carbon dioxide emissions independently from national government decisions". A criticism of the C40 network is that its exclusive nature limits influence to participating cities and risks drawing resources away from less powerful city and regional actors.

Atmospheric Dispersion

The basic technology for analyzing air pollution is through the use of a variety of mathematical models for predicting the transport of air pollutants in the lower atmosphere. The principal methodologies are:

- Point source dispersion, used for industrial sources
- Line source dispersion, used for airport and roadway air dispersion modeling
- Area source dispersion, used for forest fires or duststorms
- Photochemical models, used to analyze reactive pollutants that form smog

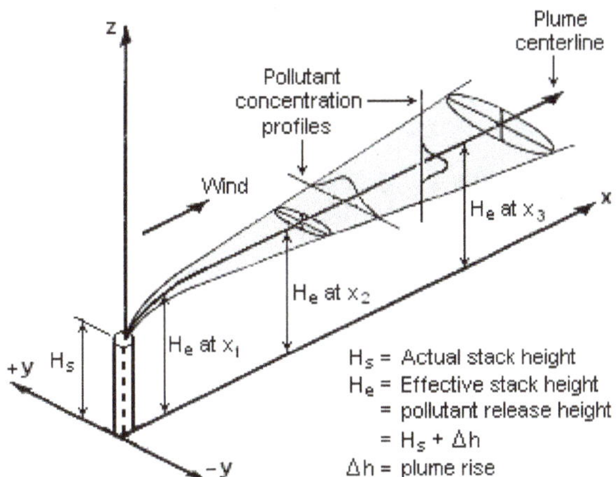

Visualization of a buoyant Gaussian air pollution dispersion plume as used in many atmospheric dispersion models.

The point source problem is the best understood, since it involves simpler mathematics and has been studied for a long period of time, dating back to about the year 1900. It uses a Gaussian dispersion model for continuous buoyant pollution plumes to predict the air pollution isopleths, with consideration given to wind velocity, stack height, emission rate and stability class (a measure of atmospheric turbulence). This model has been extensively validated and calibrated with experimental data for all sorts of atmospheric conditions.

The roadway air dispersion model was developed starting in the late 1950s and early 1960s in response to requirements of the National Environmental Policy Act and the U.S. Department of Transportation (then known as the Federal Highway Administration) to understand impacts of proposed new highways upon air quality, especially in urban areas. Several research groups were active in this model development, among which were: the Environmental Research and Technology (ERT) group in Lexington, Massachusetts, the ESL Inc. group in Sunnyvale, California and the California Air Resources Board group in Sacramento, California. The research of the ESL group received a boost with a contract award from the United States Environmental Protection Agency to validate a line source model using sulfur hexafluoride as a tracer gas. This program was successful in validating the line source model developed by ESL Inc. Some of the earliest uses of the model were in court cases involving highway air pollution; the Arlington, Virginia portion of Interstate 66 and the New Jersey Turnpike widening project through East Brunswick, New Jersey.

Area source models were developed in 1971 through 1974 by the ERT and ESL groups, but addressed a smaller fraction of total air pollution emissions, so that their use and need was not as widespread as the line source model, which enjoyed hundreds of different applications as early as the 1970s. Similarly photochemical models were developed primarily in the 1960s and 70s, but their use was more specialized and for regional needs, such as understanding smog formation in Los Angeles, California.

References

- Davis, Devra (2002). When Smoke Ran Like Water: Tales of Environmental Deception and the Battle Against Pollution. Basic Books. ISBN 0-465-01521-2.

- Turner, D.B. (1994). Workbook of atmospheric dispersion estimates: an introduction to dispersion modeling (2nd ed.). CRC Press. ISBN 1-56670-023-X.
- St. Fleur, Nicholas (10 November 2015). "Atmospheric Greenhouse Gas Levels Hit Record, Report Says". New York Times. Retrieved 11 November 2015.
- Cole, Steve; Gray, Ellen (14 December 2015). "New NASA Satellite Maps Show Human Fingerprint on Global Air Quality". NASA. Retrieved 14 December 2015.
- "Bucknell tent death: Hannah Thomas-Jones died from carbon monoxide poisoning". BBC News. 17 January 2013. Retrieved 22 September 2015.
- European Court of Justice, CURIA (2008). "PRESS RELEASE No 58/08 Judgment of the Court of Justice in Case C-237/07" (PDF). Retrieved 24 January 2015.
- House of Commons Environmental Audit Committee (2010). "Environmental Audit Committee - Fifth Report Air Quality". Retrieved 24 January 2015.
- Innes, Emma (6 June 2014). "Air pollution 'can cause changes in the brain seen in autism and schizophrenia'". Daily Mail. Retrieved 8 June 2014.
- Tankersley, Jim (January 8, 2010). "EPA proposes nation's strictest smog limits ever". Los Angeles Times. Retrieved August 14, 2012.
- Sahagun, Louis (November 13, 2008). "Pollution saps state's economy, study says". Los Angeles Times. Retrieved August 14, 2012.
- "NATA | National-Scale Air Toxics Assessments | Technology Transfer Network Air Technical Web Site | US EPA". Epa.gov. 2006-06-28. Retrieved 2012-12-11.
- European Commission. "Air quality: Commission sends final warning to UK over levels of fine particle pollution". Archived from the original on 11 May 2011. Retrieved 7 April 2011.

2

Causes of Air Pollution

A thorough account of air pollution and its causes is elucidated in this chapter. There are several causes of air pollution that have been noted, some of them being smoke, dust and volatile organic compounds. Air pollution commonly has two sources, which can be categorized as anthropogenic (man – made) sources and natural sources. This chapter elucidates the crucial theories related to air pollution.

Controlled Burn

An experimental burn in Canada; note the measuring equipment in front.

Controlled or prescribed burning, also known as hazard reduction burning (HRB), backfire, or swailing, is a technique sometimes used in forest management, farming, prairie restoration or greenhouse gas abatement. Fire is a natural part of both forest and grassland ecology and controlled fire can be a tool for foresters. Hazard reduction or controlled burning is conducted during the cooler months to reduce fuel buildup and decrease the likelihood of serious hotter fires. Controlled burning stimulates the germination of some desirable forest trees, and reveals soil mineral layers which increases seedling vitality, thus renewing the forest. Some cones, such as those of Lodgepole Pine and Sequoia, are serotinous, as well as many chaparral shrubs, meaning they require heat from fire to open cones to disperse seeds.

In industrialized countries, controlled burning is usually overseen by fire control authorities for regulations and permits. The party responsible must delineate the intended time and place. Obtaining a permit may not limit liability if the fire burns out of control.

Firing the woods in a South Carolina forest with a custom made driptorch mounted on an ATV. The device spits flaming fuel oil from the side, instantly igniting the leaf litter.

A prescribed burn in a *Pinus nigra* stand in Portugal

Controlled burning of a field outside of Statesboro, Georgia, United States in preparation for spring planting

An aerial view of a controlled burn in Helderberg Nature Reserve in South Africa bordering the city of Cape Town. In South Africa controlled burns are important for maintaining the ecological health of indigenous fynbos as well as reducing the intensity of future burns.

Japanese controlled burn (Hokkaido)

History

There are two basic causes of wildfires. One is natural (lightning) and the other is people. Controlled burns have a long history in wildland management. Pre-agricultural societies used fire to regulate both plant and animal life. Fire history studies have documented periodic wildland fires ignited by indigenous peoples in North America and Australia.

Fires, both naturally caused and prescribed, were once part of natural landscapes in many areas. Studies have shown that between the mid Holocene and the 17th century AD, wildland fires annually burned between 45.0% and 87.5% of present-day California's total land, for example. These practices ended in the early 20th century when US fire policies were enacted with the goals of suppressing all fires. Since 1995, the US Forest Service has slowly incorporated burning practices into its forest management policies.

Back Burning

Back burning is a technique utilized in controlled burning and during wildfire events. While controlled burns utilize back burning during planned fire events to create a "black line", back burning or backfiring is also done to stop a wildfire that is already in progress. Back burning is a way of reducing the amount of flammable material during a controlled burn or wildfire by starting small fires along a man made or natural firebreak in front of a main fire front. It is called back burning because the small fires are designed to 'burn back towards the main fire front' and are usually burning and traveling against ground level winds. Back burning reduces the amount of fuel that's available to the main fire by the time that it reaches the burnt area. Firebreaks are often used as an anchor point to start a line of fires along natural or manmade features such as a river, road or a bulldozed clearing.

Forest Use

Another consideration is the issue of fire prevention. In Florida, during the drought in 1995, catastrophic wildfires burned numerous homes. But forestry managers in the Florida Division of Forestry noted that the underlying problem was previous cessation of controlled burning, resulting from complaints by homeowners. Each year additional leaf litter and dropped branches increased the likelihood of a hot and uncontrollable fire.

Controlled burns are sometimes ignited using a tool known as the driptorch, which allows a steady stream of flaming fuel to be directed to the ground as needed. Variations on the driptorch can be used such as the *helitorch*, which is mounted on a helicopter, or other improvised devices such as mounting a driptorch-like device on the side of a vehicle. A pyrotechnic device known as a fusee can be used for ignition in nearby fuels while a Very pistol can be for fuels farther away.

For the burning of slash, waste materials left over from logging, there are several types of controlled burns. Broadcast burning is the burning of scattered slash over a wide area. Pile burning is gathering up the slash into piles before burning. These burning piles may be referred to as bonfires. High temperatures can harm the soil, damaging it physically, chemically or sterilizing it. Broadcast burns tend to have lower temperatures and will not harm the soil as much as pile burning, though steps can be taken to treat the soil after a burn. In lop and scatter burning, slash is left to compact over time, or is compacted with machinery. This produces a lower intensity fire, as long as the slash is not packed too tightly. However, soil may be damaged if machinery is used to compress the slash.

Controlled burning reduces fuels, may improve wildlife habitat, controls competing vegetation, improves short term forage for grazing, improves accessibility, helps control tree disease, and perpetuates fire dependent species. In mature longleaf pine forest, it helps maintain habitat for endangered red-cockaded woodpeckers in their sandhill and flatwoods habitats. Fire is also felt to be a crucial element of the recovery of the threatened Louisiana pine snake in the longleaf pine forests of central Louisiana and eastern Texas. However, many scientists disagree with such a simplistic approach, and indicate that each forest must be assessed on its own merit.

In the wild, many trees depend on fire as a successful way to clear out competition and release their seeds. In particular, the Giant sequoia depends on fire to reproduce: the cones of the tree open after a fire releases their seeds, the fire having cleared all competing vegetation. Due to fire suppression efforts during the early and mid 20th century, low-intensity fires no longer occurred naturally in many groves, and still do not occur in some groves today. The suppression of fires also led to ground fuel build-up and the dense growth which posed the risk of catastrophic wildfires. It wasn't until the 1970s that the National Park Service began systematic fires for the purpose of new seed growth. Eucalyptus regnans or Mountain Ash of Australia also depends on fire but in a different fashion. They carry their seeds in capsules which can deposit at any time of the year. Being flammable, during a fire the capsules drop nearly all of them and the fire consumes the Eucalypt adults, but most of the seeds survive using the ash as a source of nutrients; at their rate of growth, they quickly dominate the land and a new Eucalyptus forest grows.

Agricultural Use

In addition to forest management, controlled burning is also used in agriculture. In the developing world, this is often referred to as slash and burn. In industrialized nations, it is seen as one component of shifting cultivation, as a part of field preparation for planting. Often called field burning, this technique is used to clear the land of any existing crop residue as well as kill weeds and weed seeds. Field burning is less expensive than most other methods such as herbicides or tillage, but because it does produce smoke and other fire-related pollutants, its use is not popular in agricultural areas bounded by residential housing.

In the United States, field burning is a legislative and regulatory issue at both the federal and state levels of government.

Controversy

In Oregon, field burning has been widely used by grass seed farmers as a method for clearing fields for the next round of planting, as well as revitalizing serotinous grasses that require fire in order to grow seed again. The Oregon Department of Environmental Quality began requiring a permit for farmers to burn their fields in 1981, but the requirements became stricter in 1988 following a multi-car collision in which smoke from field burning near Albany, Oregon, obscured the vision of drivers on Interstate 5, leading to a 23-car collision in which 5 people died and 37 were injured. This resulted in more scrutiny of field burning and proposals to ban field burning in the state altogether.

In the European Union burning crop stubble after harvest is used by farmers for plant health reasons under several restrictions in the cross compliance regulations.

Heathland Use

The controlled burning or swailing of heathland is used in the United Kingdom and other countries as a conservation tool. In Scotland it is known as "muirburn". Often it is used as a tool for creating fire breaks to reduce the risk of dangerous outbreaks but is also an important mechanism for preventing succession to woodier vegetation and plays an important role in the life cycle of heather species. Red grouse eat heather shoots and as grouse shooting has long been an expensive sport the landowners need to ensure a good crop of grouse so each year parts of the moor are burned of old heather to cause fresh shoots.

Greenhouse Gas Abatement

Northern California fire crews start a backfire to stop the Poomacha fire from advancing westward.

Controlled burns on Australian savannas can result in an overall reduction of greenhouse gas emissions. One working example is the West Arnhem Fire Management Agreement, started to bring "strategic fire management across 28,000 square kilometres (11,000 sq mi) of Western Arnhem Land" to partially offset greenhouse gas emissions from a liquefied natural gas plant in Darwin, Australia. Deliberately starting controlled burns early in the dry season results in a mosaic of burnt and unburnt country which reduces the area of stronger, late dry season fires. Also known as "patch burning". To minimise the impact of smoke, burning should be restricted to daylight hours whenever possible.

Volatile Organic Compound

Volatile organic compounds (VOCs) are organic chemicals that have a high vapor pressure at ordinary room temperature. Their high vapor pressure results from a low boiling point, which causes large numbers of molecules to evaporate or sublimate from the liquid or solid form of the compound and enter the surrounding air, a trait known as volatility. For example, formaldehyde, which evaporates from paint, has a boiling point of only –19 °C (–2 °F).

VOCs are numerous, varied, and ubiquitous. They include both human-made and naturally occurring chemical compounds. Most scents or odors are of VOCs. VOCs play an important role in communication between plants, and messages from plants to animals. Some VOCs are dangerous to human health or cause harm to the environment. Anthropogenic VOCs are regulated by law, especially indoors, where concentrations are the highest. Harmful VOCs typically are not acutely toxic, but have compounding long-term health effects. Because the concentrations are usually low and the symptoms slow to develop, research into VOCs and their effects is difficult.

Definitions

Diverse definitions of the term VOC are in use.

The definitions of VOCs used for control of precursors of photochemical smog used by the EPA and states in the US with independent outdoor air pollution regulations include exemptions for VOCs that are determined to be non-reactive, or of low-reactivity in the smog formation process.

In the US, different regulations vary between states – most prominent is the VOC regulation by SCAQMD and by the California Air Resources Board. However, this specific use of the term VOCs can be misleading, especially when applied to indoor air quality because many chemicals that are not regulated as outdoor air pollution can still be important for indoor air pollution.

California's ARB uses the term reactive organic gases (ROG) to measure organic gases after public hearing in September 1995. The ARB revised the definition of "Volatile Organic Compounds" used in the consumer products regulations, based on their committee's findings.

Canada

Health Canada classes VOCs as organic compounds that have boiling points roughly in the range of 50 to 250 °C (122 to 482 °F). The emphasis is placed on commonly encountered VOCs that would have an effect on air quality.

European Union

A VOC is any organic compound having an initial boiling point less than or equal to 250 °C (482 °F) measured at a standard atmospheric pressure of 101.3 kPa

United States

VOCs (or specific subsets of the VOCs) are legally defined in the various laws and codes under which they are regulated. Other definitions may be found from government agencies investigating or advising about VOCs. The United States Environmental Protection Agency (EPA) regulates VOCs in the air, water, and land. The Safe Drinking Water Act implementation includes a list labeled "VOCs in connection with contaminants that are organic and volatile." The EPA also publishes testing methods for chemical compounds, some of which refer to VOCs.

In addition to drinking water, VOCs are regulated in discharges to waters (sewage treatment and stormwater disposal), as hazardous waste, but not in non industrial indoor air. The United States Department of Labor and its Occupational Safety and Health Administration (OSHA) regulate VOC exposure in the workplace. Volatile organic compounds that are hazardous material would be regulated by the Pipeline and Hazardous Materials Safety Administration while being transported.

Biologically Generated VOCs

Not counting methane, biological sources emit an estimated 1150 teragrams of carbon per year in the form of VOCs. The majority of VOCs are produced by plants, the main compound being isoprene. The remainder are produced by animals, microbes, and fungi, such as molds.

The strong odor emitted by many plants consists of green leaf volatiles, a subset of VOCs. Emissions are affected by a variety of factors, such as temperature, which determines rates of volatilization and growth, and sunlight, which determines rates of biosynthesis. Emission occurs almost exclusively from the leaves, the stomata in particular. A major class of VOCs is terpenes, such as myrcene. Providing a sense of scale, a forest 62,000 km^2 in area (the US state of Pennsylvania) is estimated to emit 3,400,000 kilograms of terpenes on a typical August day during the growing season. VOCs should be a factor in choosing which trees to plant in urban areas. Induction of genes producing volatile organic compounds, and subsequent increase in volatile terpenes has been achieved in maize using (Z)-3-Hexen-1-ol and other plant hormones.

Anthropogenic Sources

Anthropogenic sources emit about 142 teragrams of carbon per year in the form of VOCs.

Specific Components

Paints and Coatings

A major source of man-made VOCs are coatings, especially paints and protective coatings. Solvents are required to spread a protective or decorative film. Approximately 12 billion litres of paints are produced annually. Typical solvents are aliphatic hydrocarbons, ethyl acetate, glycol ethers, and acetone. Motivated by cost, environmental concerns, and regulation, the paint and coating industries are increasingly shifting toward aqueous solvents.

Chlorofluorocarbons and Chlorocarbons

Chlorofluorocarbons, which are banned or highly regulated, were widely used cleaning products and refrigerants. Tetrachloroethene is used widely in dry cleaning and by industry.

Fossil Fuels

The use of fossil fuels produces VOC's either directly as products (e.g., gasoline) or indirectly as byproducts (e.g., automobile exhaust gas).

Benzene

One VOC that is a known human carcinogen is benzene, which is a chemical found in environmental tobacco smoke, stored fuels, and exhaust from cars. Benzene also has natural sources such as volcanoes and forest fires. It is frequently used to make other chemicals in the production of plastics, resins, and synthetic fibers. Benzene evaporates into the air quickly and the vapor of benzene is heavier than air allowing the compound to sink into low-lying areas. Benzene has also been known to contaminate food and water and if digested can lead to vomiting, dizziness, sleepiness, rapid heartbeat, and at high levels, even death may occur.

Methylene Chloride

Methylene chloride can be found in adhesive removers and aerosol spray paints. In the human body, methylene chloride is metabolized to carbon monoxide. If a product that contains methylene chloride needs to be used the best way to protect human health is to use the product outdoors. If it must be used indoors, proper ventilation is will help to keep exposure levels down. In the United States, methylene chloride is listed as exempt from VOC status.

Perchloroethylene

Perchloroethylene is a volatile organic compound that has been linked to causing cancer in animals. It is also suspected to cause many of the breathing related symptoms of exposure to VOCs. Perchloroethylene is used mostly in dry cleaning. While dry cleaners recapture perchloroethylene in the dry cleaning process to reuse it, some environmental release is unavoidable.

MTBE

MTBE was banned in certain states within the US around 2004 in order to limit further contamination of drinking water aquifers (groundwater) primarily from leaking underground gasoline storage tanks where MTBE was used as an octane booster and oxygenated-additive.

Formaldehyde

Many building materials such as paints, adhesives, wall boards, and ceiling tiles slowly emit formaldehyde, which irritates the mucous membranes and can make a person irritated and uncomfortable. Formaldehyde emissions from wood are in the range of 0.02–0.04 ppm. Relative humidity within an indoor environment can also affect the emissions of formaldehyde. High relative humidity and high temperatures allow more vaporization of formaldehyde from wood-materials.

Indoor Air

Since many people spend much of their time indoors, long-term exposure to VOCs in the indoor environment can contribute to sick building syndrome. In offices, VOC results from new furnishings, wall coverings, and office equipment such as photocopy machines, which can off-gas VOCs into the air. Good ventilation and air-conditioning systems are helpful at reducing VOCs in the indoor environment. Studies also show that relative leukemia and lymphoma can increase through prolonged exposure of VOCs in the indoor environment.

In the United States, there are two standardized methods for measuring VOCs, one by the National Institute for Occupational Safety and Health (NIOSH) and another by Occupational Safety and Health Administration (OSHA). Each method uses a single component solvent; butanol and hexane cannot be sampled, however, on the same sample matrix using the NIOSH or OSHA method.

The aromatic VOC compound benzene, emitted from exhaled cigarette smoke, is labeled as carcinogenic and is ten times higher in smokers than in nonsmokers.

The United States Environmental Protection Agency (EPA) has found concentrations of VOCs in indoor air to be 2 to 5 times greater than in outdoor air and sometimes far greater. During certain activities indoor levels of VOCs may reach 1,000 times that of the outside air. Studies have shown that individual VOC emissions by themselves are not that high in an indoor environment, but the indoor total VOC (TVOC) concentrations can be up to five times higher than the VOC outdoor levels. New buildings especially, contribute to the highest level of VOC off-gassing in an indoor environment because of the abundant new materials generating VOC particles at the same time in such a short time period. In addition to new buildings, we also use many consumer products that emit VOC compounds, therefore the total concentration of VOC levels is much greater within the indoor environment.

VOC concentration in an indoor environment during winter is three to four times higher than the VOC concentrations during the summer. High indoor VOC levels are attributed to the low rates of air exchange between the indoor and outdoor environment as a result of tight-shut windows and the increasing use of humidifiers.

Indoor Air Quality Measurements

Measurement of VOCs from the indoor air is done with sorption tubes e. g. Tenax® (for VOCs and SVOCs) or DNPH-cartridges (for carbonyl-compounds). The VOCs adsorb on these materials and are afterwards desorbed either thermally (Tenax®) or by elution (DNPH) and then analyzed by GC-MS/FID or HPLC. Reference gas mixtures are required for quality control of these VOC-measurements. Furthermore, VOC emitting products used indoors, e. g. building products and furniture, are investigated in emission test chambers under controlled climatic conditions. For quality control of these measurements round robin tests are carried out, therefore reproducibly emitting reference materials are ideally required.

Regulation of Indoor VOC Emissions

In most countries, a separate definition of VOCs is used with regard to indoor air quality that comprises each organic chemical compound that can be measured as follows: Adsorption from air on Tenax TA, thermal desorption, gas chromatographic separation over a 100% nonpolar column (dimethylpolysiloxane). VOC (volatile organic compounds) are all compounds that appear in the gas chromatogram between and including n-hexane and n-hexadecane. Compounds appearing earlier are called VVOC (very volatile organic compounds) compounds appearing later are called SVOC (semi-volatile organic compounds).

France, Germany and Belgium have enacted regulations to limit VOC emissions from commercial products, and industry has developed numerous voluntary ecolabels and rating systems, such as EMICODE, M1, Blue Angel and Indoor Air Comfort In the United States, several standards exist; California Standard CDPH Section 01350 is the most popular one. Over the last few decades, these regulations and standards changed the marketplace, leading to an increasing number of low-emitting products. The leading voluntary labels report that licenses to several hundreds of low-emitting products have been issued.

Health Risks

Respiratory, allergic, or immune effects in infants or children are associated with man-made VOCs and other indoor or outdoor air pollutants.

Some VOCs, such as styrene and limonene, can react with nitrogen oxides or with ozone to produce new oxidation products and secondary aerosols, which can cause sensory irritation symptoms. Unspecified VOCs are important in the creation of smog.

Health effects include eye, nose, and throat irritation; headaches, loss of coordination, nausea; and damage to the liver, kidney, and central nervous system. Some organics can cause cancer in animals; some are suspected or known to cause cancer in humans. Key signs or symptoms associated with exposure to VOCs include conjunctival irritation, nose and throat discomfort, headache, allergic skin reaction, dyspnea, declines in serum cholinesterase levels, nausea, vomiting, nose bleeding, fatigue, dizziness.

The ability of organic chemicals to cause health effects varies greatly from those that are highly toxic, to those with no known health effects. As with other pollutants, the extent and nature of the health effect will depend on many factors including level of exposure and length of time exposed. Eye and respiratory tract irritation, headaches, dizziness, visual disorders, and memory impairment are among the immediate symptoms that some people have experienced soon after exposure to some organics. At present, not much is known about what health effects occur from the levels of organics usually found in homes. Many organic compounds are known to cause cancer in animals; some are suspected of causing, or are known to cause, cancer in humans.

Reducing Exposure

To reduce exposure to these toxins, one should buy products that contain Low-VOCs or No VOCs. Only the quantity which will soon be needed should be purchased, eliminating stockpiling of these chemicals. Use products with VOCs in well ventilated areas. When designing homes and buildings, design teams can implement the best possible ventilation plans, call for the best mechanical systems available, and design assemblies to reduce the amount of infiltration into the building. These methods will help improve indoor air quality, but by themselves they cannot keep a building from becoming an unhealthy place to breathe.

Limit Values for VOC Emissions

Limit values for VOC emissions into indoor air are published by e.g. AgBB, AFSSET, California Department of Public Health, and others. These regulations have prompted several companies to adapt with VOC level reductions in products that have VOCs in their formula, such as BEHR, KILZ, and Benjamin Moore & Co. in the paint industry and Weld-On in the adhesive industry.

Chemical Fingerprinting

The exhaled human breath contains a few hundred volatile organic compounds and is used in breath analysis to serve as a VOC biomarker to test for diseases such as lung cancer. One study has shown that "volatile organic compounds ... are mainly blood borne and therefore enable monitoring of different processes in the body." And it appears that VOC compounds in the body "may be either produced by metabolic processes or inhaled/absorbed from exogenous sources" such as environmental tobacco smoke. Research is still in the process to determine whether VOCs in the body are contributed by cellular processes or by the cancerous tumors in the lung or other organs.

VOC Sensors

Principle and Measurement Methods

VOCs in the environment or certain atmospheres can be detected based on different principles and interactions between the organic compounds and the sensor components. There are electronic devices that can detect ppm concentrations despite the non-selectivity. Others can predict with reasonable accuracy the molecular structure of the volatile organic compounds in the environment or enclosed atmospheres and could be used as accurate monitors of the Chemical Fingerprint and further as health monitoring devices.

Solid-phase microextraction (SPME) techniques are used to collect VOCs at low concentrations for analysis.

Direct injection mass spectrometry techniques are frequently utilized for the rapid detection and accurate quantification of VOCs. PTR-MS is among the methods that have been used most extensively for the on-line analysis of biogenic and antropogenic VOCs. Recent PTR-MS instruments based on time-of-flight mass spectrometry have been reported to reach detection limits of 20 pptv after 100 ms and 750 ppqv after 1 min. measurement (signal integration) time. The mass resolution of these devices is between 7000 and 10,500 $m/\Delta m$, thus it is possible to separate most common isobaric VOCs and quantify them independently.

Accuracy and Traceability

Metrology for VOC Measurements

To achieve comparability of VOC measurements, reference standards traceable to SI-units are required. For a number of VOCs gaseous reference standards are available from specialty gas suppliers or national metrology institutes, either in the form of cylinders or dynamic generation methods. However, for many VOCs, such as oxygenated VOCs, monoterpenes, or formaldehyde, no standards are available at the appropriate amount of fraction due to the chemical reactivity or adsorption of these molecules. Currently, several national metrology institutes are working on the lacking standard gas mixtures at trace level concentration, minimising adsorption processes, and improving the zero gas. The final scopes are for the traceability and the long-term stability of the standard gases to be in accordance with the data quality objectives (DQO, maximum uncertainty of 20% in this case) required by the WMO/GAW program.

Smoke

Smoke is a collection of airborne solid and liquid particulates and gases emitted when a material undergoes combustion or pyrolysis, together with the quantity of air that is entrained or otherwise mixed into the mass. It is commonly an unwanted by-product of fires (including stoves, candles, oil lamps, and fireplaces), but may also be used for pest control (fumigation), communication (smoke signals), defensive and offensive capabilities in the military (smoke-screen), cooking, or smoking (tobacco, cannabis, etc.). Smoke is used in rituals where incense, sage, or resin is burned to produce a smell for spiritual purposes. Smoke is sometimes used as a flavoring agent, and preservative for various foodstuffs. Smoke is also a component of internal combustion engine exhaust gas, particularly diesel exhaust.

Smoke

Smoke from a bee smoker, used in beekeeping.

Smoke inhalation is the primary cause of death in victims of indoor fires. The smoke kills by a combination of thermal damage, poisoning and pulmonary irritation caused by carbon monoxide, hydrogen cyanide and other combustion products.

Smoke is an aerosol (or mist) of solid particles and liquid droplets that are close to the ideal range of sizes for Mie scattering of visible light. This effect has been likened to three-dimensional textured privacy glass — a smoke cloud does not obstruct an image, but thoroughly scrambles it.

Chemical Composition

The composition of smoke depends on the nature of the burning fuel and the conditions of combustion.

Fires with high availability of oxygen burn at a high temperature and with small amount of smoke produced; the particles are mostly composed of ash, or with large temperature differences, of condensed aerosol of water. High temperature also leads to production of nitrogen oxides. Sulfur content yields sulfur dioxide, or in case of incomplete combustion, hydrogen sulfide. Carbon and hydrogen are almost completely oxidized to carbon dioxide and water. Fires burning with lack of oxygen produce a significantly wider palette of compounds, many of them toxic. Partial oxidation of carbon produces carbon monoxide, nitrogen-containing materials can yield hydrogen cyanide, ammonia, and nitrogen oxides. Hydrogen gas can be produced instead of water. Content of halogens such as chlorine (e.g. in polyvinyl chloride or brominated flame retardants) may lead to production of e.g. hydrogen chloride, phosgene, dioxin, and chloromethane, bromomethane and other halocarbons. Hydrogen fluoride can be formed from fluorocarbons, whether fluoropolymers subjected to fire or halocarbon fire suppression agents. Phosphorus and antimony oxides and their reaction products can be formed from some fire retardant additives, increasing smoke toxicity and corrosivity. Pyrolysis of polychlorinated biphenyls (PCB), e.g. from burning older transformer oil, and to lower degree also of other chlorine-containing materials, can produce 2,3,7,8-tetrachlorodibenzodioxin, a potent carcinogen, and other polychlorinated dibenzodioxins. Pyrolysis of fluoropolymers, e.g. teflon, in presence of oxygen yields carbonyl fluoride (which hydrolyzes readily to HF and CO_2); other compounds may be formed as well, e.g. carbon tetrafluoride, hexafluoropropylene, and highly toxic perfluoroisobutene (PFIB).

Emission of soot in the fumes of a large diesel truck, without particle filters.

Pyrolysis of burning material, especially incomplete combustion or smoldering without adequate oxygen supply, also results in production of a large amount of hydrocarbons, both aliphatic (methane, ethane, ethylene, acetylene) and aromatic (benzene and its derivates, polycyclic aromatic hydrocarbons; e.g. benzo[a]pyrene, studied as a carcinogen, or retene), terpenes. Heterocyclic compounds may be also present. Heavier hydrocarbons may condense as tar; smoke with significant tar content is yellow to brown. Presence of such smoke, soot, and/or brown oily deposits during a fire indicates a possible hazardous situation, as the atmosphere may be saturated with combustible pyrolysis products with concentration above the upper flammability limit, and sudden inrush of air can cause flashover or backdraft.

Presence of sulfur can lead to formation of e.g. hydrogen sulfide, carbonyl sulfide, sulfur dioxide, carbon disulfide, and thiols; especially thiols tend to get adsorbed on surfaces and produce a lingering odor even long after the fire. Partial oxidation of the released hydrocarbons yields in a wide palette of other compounds: aldehydes (e.g. formaldehyde, acrolein, and furfural), ketones, alcohols (often aromatic, e.g. phenol, guaiacol, syringol, catechol, and cresols), carboxylic acids (formic acid, acetic acid, etc.).

The visible particulate matter in such smokes is most commonly composed of carbon (soot). Other particulates may be composed of drops of condensed tar, or solid particles of ash. The presence of metals in the fuel yields particles of metal oxides. Particles of inorganic salts may also be formed, e.g. ammonium sulfate, ammonium nitrate, or sodium chloride. Inorganic salts present on the surface of the soot particles may make them hydrophilic. Many organic compounds, typically the aromatic hydrocarbons, may be also adsorbed on the surface of the solid particles. Metal oxides can be present when metal-containing fuels are burned, e.g. solid rocket fuels containing aluminium. Depleted uranium projectiles after impacting the target ignite, producing particles of uranium oxides. Magnetic particles, spherules of magnetite-like ferrous ferric oxide, are present in coal smoke; their increase in deposits after 1860 marks the beginning of the Industrial Revolution. (Magnetic iron oxide nanoparticles can be also produced in the smoke from meteorites burning in the atmosphere.) Magnetic remanence, recorded in the iron oxide particles, indicates the strength of Earth's magnetic field when they were cooled beyond their Curie temperature; this can be used to distinguish magnetic particles of terrestrial and meteoric origin. Fly ash is composed mainly of silica and calcium oxide. Cenospheres are present in smoke from liquid hydrocarbon fuels. Minute metal particles produced by abrasion can be present in engine smokes. Amorphous silica particles are present in smokes from burning silicones; small proportion of silicon nitride particles can be formed in fires with insufficient oxygen. The silica particles have about 10 nm size, clumped to 70-100 nm aggregates and further agglomerated to chains. Radioactive particles may be present due to traces of uranium, thorium, or other radionuclides in the fuel; hot particles can be present in case of fires during nuclear accidents (e.g. Chernobyl disaster) or nuclear war.

Smoke particulates, like other aerosols, are categorized into three modes based on particle size:

- nuclei mode, with geometric mean radius between 2.5–20 nm, likely forming by condensation of carbon moieties.
- accumulation mode, ranging between 75–250 nm and formed by coagulation of nuclei mode particles
- coarse mode, with particles in micrometer range

Most of the smoke material is primarily in coarse particles. Those undergo rapid dry precipitation, and the smoke damage in more distant areas outside of the room where the fire occurs is therefore primarily mediated by the smaller particles.

Aerosol of particles beyond visible size is an early indicator of materials in a preignition stage of a fire.

Burning of hydrogen-rich fuel produces water; this results in smoke containing droplets of water vapor. In absence of other color sources (nitrogen oxides, particulates...), such smoke is white and cloud-like.

Smoke emissions may contain characteristic trace elements. Vanadium is present in emissions from oil fired power plants and refineries; oil plants also emit some nickel. Coal combustion produces emissions containing aluminium, arsenic, chromium, cobalt, copper, iron, mercury, selenium, and uranium.

Traces of vanadium in high-temperature combustion products form droplets of molten vanadates. These attack the passivation layers on metals and cause high temperature corrosion, which is a concern especially for internal combustion engines. Molten sulfate and lead particulates also have such effect.

Some components of smoke are characteristic of the combustion source. Guaiacol and its derivatives are products of pyrolysis of lignin and are characteristic of wood smoke; other markers are syringol and derivates, and other methoxy phenols. Retene, a product of pyrolysis of conifer trees, is an indicator of forest fires. Levoglucosan is a pyrolysis product of cellulose. Hardwood vs softwood smokes differ in the ratio of guaiacols/syringols. Markers for vehicle exhaust include polycyclic aromatic hydrocarbons, hopanes, steranes, and specific nitroarenes (e.g. 1-nitropyrene). The ratio of hopanes and steranes to elemental carbon can be used to distinguish between emissions of gasoline and diesel engines.

Many compounds can be associated with particulates; whether by being adsorbed on their surfaces, or by being dissolved in liquid droplets. Hydrogen chloride is well absorbed in the soot particles.

Inert particulate matter can be disturbed and entrained into the smoke. Of particular concern are particles of asbestos.

Deposited hot particles of radioactive fallout and bioaccumulated radioisotopes can be reintroduced into the atmosphere by wildfires and forest fires; this is a concern in e.g. the Zone of alienation containing contaminants from the Chernobyl disaster.

Polymers are a significant source of smoke. Aromatic side groups, e.g. in polystyrene, enhance generation of smoke. Aromatic groups integrated in the polymer backbone produce less smoke, likely due to significant charring. Aliphatic polymers tend to generate the least smoke, and are non-self-extinguishing. However presence of additives can significantly increase smoke formation. Phosphorus-based and halogen-based flame retardants decrease production of smoke. Higher degree of cross-linking between the polymer chains has such effect too.

Visible and Invisible Particles of Combustion

The naked eye detects particle sizes greater than 7 µm (micrometres). Visible particles emitted from a fire are referred to as smoke. Invisible particles are generally referred to as gas or fumes. This is best illustrated when toasting bread in a toaster. As the bread heats up, the products of combustion increase in size. The fumes initially produced are invisible but become visible if the toast is burnt.

Smoke from a wildfire

Smoke rising up from the smoldering remains of a recently extingished mountain fire in South Africa.

An ionization chamber type smoke detector is technically a product of combustion detector, not a smoke detector. Ionization chamber type smoke detectors detect particles of combustion that are invisible to the naked eye. This explains why they may frequently false alarm from the fumes emitted from the red-hot heating elements of a toaster, before the presence of visible smoke, yet they may fail to activate in the early, low-heat smoldering stage of a fire.

Smoke from a typical house fire contains hundreds of different chemicals and fumes. As a result, the damage caused by the smoke can often exceed that caused by the actual heat of the fire. In addition to the physical damage caused by the smoke of a fire – which manifests itself in the form of stains – is the often even harder to eliminate problem of a smoky odor. Just as there are contractors that specialize in rebuilding/repairing homes that have been damaged by fire and smoke, fabric restoration companies specialize in restoring fabrics that have been damaged in a fire.

Dangers of Smoke

Smoke from oxygen-deprived fires contains a significant concentration of compounds that are flammable. A cloud of smoke, in contact with atmospheric oxygen, therefore has the potential of being ignited – either by another open flame in the area, or by its own temperature. This leads to effects like backdraft and flashover. Smoke inhalation is also a danger of smoke that can cause serious injury and death.

Many compounds of smoke from fires are highly toxic and/or irritating. The most dangerous is carbon monoxide leading to carbon monoxide poisoning, sometimes with the additive effects of hydrogen cyanide and phosgene. Smoke inhalation can therefore quickly lead to incapacitation and loss of consciousness. Sulfur oxides, hydrogen chloride and hydrogen fluoride in contact with moisture form sulfuric, hydrochloric and hydrofluoric acid, which are corrosive to both lungs and materials. When asleep the nose does not sense smoke nor does the brain, but the body will wake up if the lungs become enveloped in smoke and the brain will be stimulated and the person will be awoken. This does not work if the person is incapacitated or under the influence of drugs and/or alcohol.

Cigarette smoke is a major modifiable risk factor for lung disease, heart disease, and many cancers. Smoke can also be a component of ambient air pollution due to the burning of coal in power plants, forest fires or other sources, although the concentration of pollutants in ambient air is typically much less than that in cigarette smoke. One day of exposure to PM2.5 at a concentration of 880 μg/m3, such as occurs in Beijing, China, is the equivalent of smoking one or two cigarettes in terms of particulate inhalation by weight. The analysis is complicated, however, by the fact that the organic compounds present in various ambient particulates may have a higher carcinogenicity than the compounds in cigarette smoke particulates.

Reduced visibility due to wildfire smoke in Sheremetyevo airport (Moscow, Russia) 7 August 2010.

Smoke can obscure visibility, impeding occupant exiting from fire areas. In fact, the poor visibility due to the smoke that was in the Worcester Cold Storage Warehouse fire in Worcester, Massachusetts was the exact reason why the trapped rescue firefighters couldn't evacuate the building in time. Because of the striking similarity that each floor shared, the dense smoke caused the firefighters to become disoriented.

The effect of smoke burning off the end of a tobacco product or the smoke exhaled by a smoker is known as secondhand smoke. This smoke is contained with harmful substances that can harm the body of the smoker. A person could have serious health problems from the result of breathing in secondhand smoke such as diseases or cancers. A child's undeveloped body can face respiratory problems that could affect their lives forever. Secondhand smoke also harms the environment by being an indoor air pollutant that people breathe in.

Smoke Corrosion

Smoke contains a wide variety of chemicals, many of them aggressive in nature. Examples are hydrochloric acid and hydrobromic acid, produced from halogen-containing plastics and fire retardants, hydrofluoric acid released by pyrolysis of fluorocarbon fire suppression agents, sulfuric acid from burning of sulfur-containing materials, nitric acid from high-temperature fires where nitrous oxide gets formed, phosphoric acid and antimony compounds from P and Sb based fire retardants, and many others. Such corrosion is not significant for structural materials, but delicate structures, especially microelectronics, are strongly affected. Corrosion of circuit board traces, penetration of aggressive chemicals through the casings of parts, and other effects can cause an immediate or gradual deterioration of parameters or even premature (and often delayed, as the corrosion can progress over long time) failure of equipment subjected to smoke. Many smoke components are also electrically conductive; deposition of a conductive layer on the circuits can cause crosstalks and other deteriorations of the operating parameters or even cause short circuits and total failures. Electrical contacts can be affected by corrosion of surfaces, and by deposition of soot and other conductive particles or nonconductive layers on or across the contacts. Deposited particles may adversely affect the performance of optoelectronics by absorbing or scattering the light beams.

Corrosivity of smoke produced by materials is characterized by the corrosion index (CI), defined as material loss rate (angstrom/minute) per amount of material gasified products (grams) per volume of air (m^3). It is measured by exposing strips of metal to flow of combustion products in a test tunnel. Polymers containing halogen and hydrogen (polyvinyl chloride, polyolefins with halogenated additives, etc.) have the highest CI as the corrosive acids are formed directly with water produced by the combustion, polymers containing halogen only (e.g. polytetrafluoroethylene) have lower CI as the formation of acid is limited to reactions with airborne humidity, and halogen-free materials (polyolefins, wood) have the lowest CI. However, some halogen-free materials can also release significant amount of corrosive products.

Smoke damage to electronic equipment can be significantly more extensive than the fire itself. Cable fires are of special concern; low smoke zero halogen materials are preferable for cable insulation.

When smoke comes into contact with the surface of any substance or structure, the chemicals contained in it are transferred to it. The corrosive properties of the chemicals cause the substance or structure to decompose at a rapid rate. Certain materials or structures absorb these chemicals, which is why clothing, unsealed surfaces, potable water, piping, wood, etc., are replaced in most cases of structural fires.

Secondhand Tobacco Smoke Inhalation

Secondhand tobacco smoke is the combination of both sidestream and mainstream smoke emissions. These emissions contain more than 50 carcinogenic chemicals. According to the Surgeon

General's latest report on the subject, "Short exposures to secondhand [tobacco] smoke can cause blood platelets to become stickier, damage the lining of blood vessels, decrease coronary flow velocity reserves, and reduce heart variability, potentially increasing the risk of a heart attack". The American Cancer Society lists "heart disease, lung infections, increased asthma attacks, middle ear infections, and low birth weight" as ramifications of smoker's emission.

Measurement

As early as the 15th century Leonardo da Vinci commented at length on the difficulty of assessing smoke, and distinguished between black smoke (carbonized particles) and white 'smoke' which is not a smoke at all but merely a suspension of harmless water particulates. Smoke from heating appliances is commonly measured in one of the following ways:

In-line capture. A smoke sample is simply sucked through a filter which is weighed before and after the test and the mass of smoke found. This is the simplest and probably the most accurate method, but can only be used where the smoke concentration is slight, as the filter can quickly become blocked.

The ASTM smoke pump is a simple and widely used method of in-line capture where a measured volume of smoke is pulled through a filter paper and the dark spot so formed is compared with a standard.

Filter/dilution tunnel. A smoke sample is drawn through a tube where it is diluted with air, the resulting smoke/air mixture is then pulled through a filter and weighed. This is the internationally recognized method of measuring smoke from combustion.

Electrostatic precipitation. The smoke is passed through an array of metal tubes which contain suspended wires. A (huge) electrical potential is applied across the tubes and wires so that the smoke particles become charged and are attracted to the sides of the tubes. This method can over-read by capturing harmless condensates, or under-read due to the insulating effect of the smoke. However, it is the necessary method for assessing volumes of smoke too great to be forced through a filter, i.e., from bituminous coal.

Ringelmann scale. A measure of smoke color. Invented by Professor Maximilian Ringelmann in Paris in 1888, it is essentially a card with squares of black, white and shades of gray which is held up and the comparative grayness of the smoke judged. Highly dependent on light conditions and the skill of the observer it allocates a grayness number from 0 (white) to 5 (black) which has only a passing relationship to the actual quantity of smoke. Nonetheless, the simplicity of the Ringelmann scale means that it has been adopted as a standard in many countries.

Optical scattering. A light beam is passed through the smoke. A light detector is situated at an angle to the light source, typically at 90°, so that it receives only light reflected from passing particles. A measurement is made of the light received which will be higher as the concentration of smoke particles becomes higher.

Optical obscuration. A light beam is passed through the smoke and a detector opposite measures the light. The more smoke particles are present between the two, the less light will be measured.

Combined optical methods. There are various proprietary optical smoke measurement devices such as the 'nephelometer' or the 'aethalometer' which use several different optical methods, including more than one wavelength of light, inside a single instrument and apply an algorithm to give a good estimate of smoke. It has been claimed that these devices can differentiate types of smoke and so their probable source can be inferred, though this is disputed.

Inference from carbon monoxide. Smoke is incompletely burned fuel, carbon monoxide is incompletely burned carbon, therefore it has long been assumed that measurement of CO in flue gas (a cheap, simple and very accurate procedure) will provide a good indication of the levels of smoke. Indeed, several jurisdictions use CO measurement as the basis of smoke control. However it is far from clear how accurate the correspondence is.

Medicinal Smoke

Throughout recorded history, humans have used the smoke of medicinal plants to cure illness. A sculpture from Persepolis shows Darius the Great (522–486 BC), the king of Persia, with two censers in front of him for burning Peganum harmala and/or sandalwood Santalum album, which was believed to protect the king from evil and disease. More than 300 plant species in 5 continents are used in smoke form for different diseases. As a method of drug administration, smoking is important as it is a simple, inexpensive, but very effective method of extracting particles containing active agents. More importantly, generating smoke reduces the particle size to a microscopic scale thereby increasing the absorption of its active chemical principles.

Dust

Dust consists of particles in the atmosphere that come from various sources such as soil, dust lifted by weather (an aeolian processes), volcanic eruptions, and pollution. Dust in homes, offices, and other human environments contains small amounts of plant pollen, human and animal hairs, textile fibers, paper fibers, minerals from outdoor soil, human skin cells, burnt meteorite particles, and many other materials which may be found in the local environment.

A dust storm blankets Texas homes, 1935

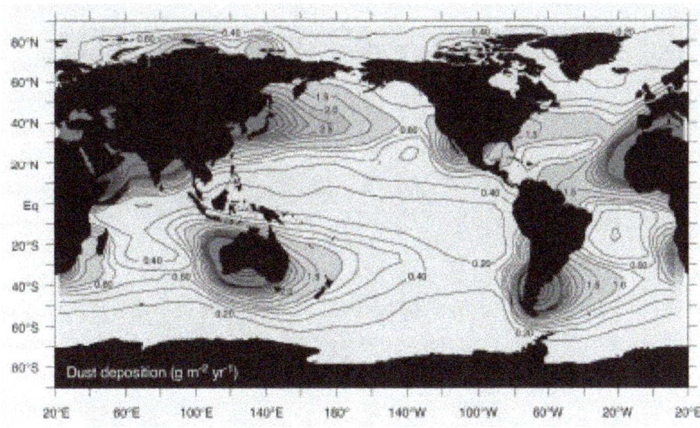

Global oceanic distribution of dust deposition

Domestic Dust and Humans

Three years of use without cleaning has caused this laptop heat sink to become clogged with dust, and it can no longer be used as it may overheat.

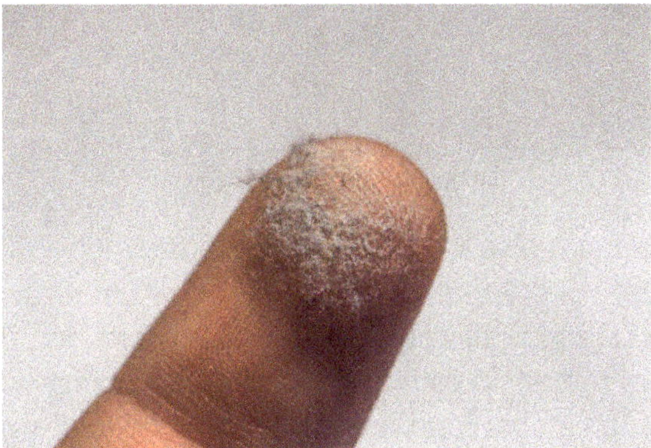

domestic dust on a finger

House dust mites are present indoors wherever humans live. Positive tests for dust mite allergies are extremely common among people with asthma. Dust mites are microscopic arachnids whose primary food is dead human skin cells, but they do not live on living people. They and their feces and other allergens they produce are major constituents of house dust, but because they are so heavy they are not suspended for long in the air. They are generally found on the floor and other surfaces until disturbed (by walking, for example). It could take somewhere between twenty minutes and two hours for dust mites to settle back down out of the air.

Dust mites are a nesting species that prefers a dark, warm, and humid climate. They flourish in mattresses, bedding, upholstered furniture, and carpets. Their feces include enzymes that are released upon contact with a moist surface, which can happen when a person inhales, and these enzymes can kill cells within the human body. House dust mites did not become a problem until humans began to use textiles, such as western style blankets and clothing.

Atmospheric Dust

Presentation on imported dust in North American skies

Large dust storm over Libya

Atmospheric or wind-borne dust, also known as *aeolian dust*, comes from arid and dry regions where high velocity winds are able to remove mostly silt-sized material, deflating susceptible surfaces. This includes areas where grazing, ploughing, vehicle use, and other human activities have further destabilized the land, though not all source areas have been largely affected by anthropogenic impacts. One-third of the global land area is covered by dust-producing surfaces, made up of hyper-arid regions like the Sahara which covers 0.9 billion hectares, and drylands which occupy 5.2 billion hectares.

Dust in the atmosphere is produced by saltation and sandblasting of sand-sized grains, and it is transported through the troposphere. This airborne dust is considered an aerosol and once in the atmosphere, it can produce strong local radiative forcing. Saharan dust in particular can be transported and deposited as far as the Caribbean and Amazonia, and may affect air temperatures, cause ocean cooling, and alter rainfall amounts.

Middle East

Dust in the Middle East has been a historic phenomenon. Recently, because of climate change and the escalating process of desertification, the problem has worsened dramatically. As a multi-factor phenomenon, there is not yet a clear consensus on the sources or potential solutions to the problem.

In Iran, the dust is already affecting more than 5 million people directly, and has emerged as a serious government issue in recent years. In the province of Khuzestan it has led to the severe reduction of air quality. The amount of pollutants in the air has surpassed more than 50 times the normal level several times in a year. Recently, initiatives such as Project-Dust have been established to directly study the Middle Eastern dust.

Road Dust

Dust kicked up by vehicles traveling on roads may make up 33% of air pollution. Road dust consists of deposits of vehicle exhausts and industrial exhausts, particles from tire and brake wear, dust from paved roads or potholes, and dust from construction sites. Road dust is a significant source contributing to the generation and release of particulate matter into the atmosphere. Control of road dust is a significant challenge in urban areas, and also in other locations with high levels of vehicular traffic upon unsealed roads, such as mines and landfill dumps.

Road dust may be suppressed by mechanical methods like street sweeper vehicles equipped with vacuum cleaners, vegetable oil sprays, or with water sprayers. Improvements in automotive engineering have reduced the amount of PM10s produced by road traffic; the proportion representing re-suspension of existing particulates has, increased as a result.

Coal Dust

Coal dust is responsible for the lung disease known as pneumoconiosis, including black lung disease that occurs among coal miners. The danger of coal dust resulted in environmental legislation regulating work place air quality in some jurisdictions. In addition, if enough coal dust is dispersed within the air in a given area, in very rare circumstances, it can create an explosion hazard under certain circumstances. These circumstances are typically within confined spaces.

Dust Control

Control of Atmospheric Dust

Most governmental EPAs, including the United States Environmental Protection Agency (EPA) mandate that facilities that generate dust, minimize or mitigate the production of dust in their operation. The most frequent dust control violations occur at new residential housing developments in urban areas. United States Federal law requires that construction sites obtain permits to conduct earth moving, clearing of areas, to include plans to control dust emissions when the work is being carried out. Control measures include such simple practices as spraying construction and demolition sites with water, and preventing the tracking of dust onto adjacent roads.

Some of the issues include:

- Reducing dust related health risks that include allergic reactions, pneumonia and asthmatic attacks.
- Improving visibility and road safety.

- Providing cleaner air, cleaner vehicles and cleaner homes and promoting better health.
- Improving crop productivity in agriculture.
- Reducing vehicle maintenance costs by lowering the levels of dust that clog filters, bearings and machinery.
- Reducing driver fatigue, maintenance on suspension systems and improving fuel economy.
- Increasing cumulative effect - each new application builds on previous residuals reducing re-application rate *while improving performance.

US federal laws require dust control on sources such as vacant lots, unpaved parking lots, and unpaved roads. Dust in such places may be suppressed by mechanical methods, including paving or laying down gravel, or stabilizing the surface with water, vegetable oils or other dust suppressants, or by using water misters to suppress dust that is already airborne.

Control of Domestic Dust

Dust on the air filter

Dust control is the suppression of solid particles with diameters less than 500 micrometers. Dust poses a health threat to children, older people, and those with respiratory illnesses.

House dust can become airborne easily. Care is required when removing dust to avoid causing the dust to become airborne. A feather duster tends to agitate the dust so it lands elsewhere. Products like Pledge and Swiffer are specifically made for removing dust by trapping it with sticky chemicals.

Certified HEPA (tested to MIL STD 282) can effectively trap 99.97% of dust at 0.3 micrometers. Not all HEPA (type/media) filters can effectively stop dust; while vacuum cleaners with HEPA (type/media) filters, water, or cyclones may filter more effectively than without, they may still exhaust millions of particles per cubic foot of air circulated. Central vacuum cleaners can be effective in removing dust, especially if they are exhausted directly to the outdoors.

Air filtering appliances differ greatly in their effectiveness. Laser particle counters are an effective way to measure filter effectiveness, medical grade instruments can test for particles as small as 0.3 micrometers. In order to test for dust in the air, there are several options available. Pre weighted filter and matched weight filters made from polyvinyl chloride or mixed cellulose ester are suitable for respirable dust (less than 10 micrometers in diameter).

Control of Dust Resistance on Surfaces

A dust resistant surface is a state of prevention against dust contamination or damage, by a design or treatment of materials and items in manufacturing or through a repair process. A reduced tacticity of a synthetic layer or covering can protect surfaces and release small molecules that could have remained attached. A panel, container or enclosure with seams may feature types of strengthened rigidity or sealant to vulnerable edges and joins.

Dust in Outer Space

Cosmic dust is widely present in space, where gas and dust clouds are primary precursors for planetary systems. The zodiacal light, as seen in a dark night sky, is produced by sunlight reflected from particles of dust in orbit around the Sun. The tails of comets are produced by emissions of dust and ionized gas from the body of the comet. Dust also covers solid planetary bodies, and vast dust storms occur on Mars that cover almost the entire planet. Interstellar dust is found between the stars, and high concentrations produce diffuse nebulae and reflection nebulae.

Dust is widely present in the galaxy. Ambient radiation heats dust and re-emits radiation into the microwave band, which may distort the cosmic microwave background power spectrum. Dust in this regime has a complicated emission spectrum, and includes both thermal dust emission and spinning dust emission.

Dust samples returned from outer space may provide information about conditions in the early solar system. Several spacecraft have sought to gather samples of dust and other materials. Among these craft was Stardust, which flew past Comet Wild 2 in 2004, and returned a capsule of the comet's remains to Earth in January 2006. In 2010 the Japanese Hayabusa spacecraft returned samples of dust from the surface of an asteroid.

Examples of Atmospheric Dust

Dry, windy weather sends clouds of dust across south-eastern Australia.

A pale brown plume of dust sweeps out of Argentina's Pampas.

A thick dust plume over Kuwait and the north-western tip of the Persian Gulf.

References

- Stoye, D.; Funke, W.; Hoppe, L.; et al. (2006), "Paints and Coatings", Ullmann's Encyclopedia of Industrial Chemistry, doi:10.1002/14356007.a18_359.pub2, ISBN 3527306730
- Ellis, Andrew M.; Mayhew, Christopher A. (2014). Proton Transfer Reaction Mass Spectrometry - Principles and Applications. Chichester, West Sussex, UK: John Wiley & Sons Ltd. ISBN 978-1-405-17668-2.
- Carlone, Nancy (2009). Nancy Caroline's Emergency Care in the Streets, Canadian Edition. Burlington, Massachusetts: Jones & Bartlett Learning. pp. 20–28. ISBN 9781284053845. Retrieved 22 July 2014.
- Mauseth, James D. (1991). Botany: An Introduction to Plant Biology. Burlington, Massachusetts: Jones & Bartlett Learning. p. 234. ISBN 9780030938931. Retrieved 22 July 2014. .
- Reuter, M.A.; Boin, U.M.J.; Schaik, A. van; Verhoef, E.; Heiskanen, K.; Yang, Yongxiang; Georgalli, G. (2 November 2005). The Metrics of Material and Metal Ecology. Amsterdam: Elsevier. ISBN 9780080457925. Retrieved 22 July 2014.
- Moldoveanu, Serban (16 September 2009). Pyrolysis of Organic Molecules: Applications to Health and Environmental Issues. Elsevier. p. 643. ISBN 0444531130. Retrieved 2014-11-20.
- Fire, Frank L. (2009). The Common Sense Approach to Hazardous Materials. Fire Engineering Books. p. 129. ISBN 978-0912212111. Retrieved 2014-11-20.
- Krevelen, D.W. van; Nijenhuis, Klaas te (2009). Properties of Polymers: Their Correlation with Chemical Structure; Their Numerical Estimation and Prediction from Additive Group Contributions. Elsevier. p. 864. ISBN 0-08-054819-9.
- Ronald C. Lasky, Ronald Lasky, Ulf L. Österberg, Daniel P. Stigliani (1995). Optoelectronics for data communication. Academic Press. p. 43. ISBN 0-12-437160-4. CS1 maint: Uses authors parameter (link)

- Watson, Donna S. (8 March 2010). Perioperative Safety. Amsterdam, Netherlands: Elsevier Health Sciences. ISBN 978-0-323-06985-4. Retrieved 23 August 2014.
- Harrison & others, Roy M (26 August 2013). "An evaluation of some issues regarding the use of aethalometers to measure woodsmoke concentrations". Atmospheric Environment. 80: 540–548. Bibcode:2013AtmEn..80..540H. doi:10.1016/j.atmosenv.2013.08.026. Retrieved 20 March 2015.
- R.A. Bradstock; M. Bedward; B.J. Kenny; J. Scott (1998). "Spatially-explicit simulation of the effect of prescribed burning on fire regimes and plant extinctions in shrublands typical of south-eastern Australia". Biological Conservation 86 (1): 83–95. doi:10.1016/S0006-3207(97)00170-5. Retrieved 28 February 2012.
- Scott L. Stephens; Robert E. Martin; Nicholas E. Clinton (2007). "Prehistoric fire area and emissions from California's forests, woodlands, shrublands, and grasslands". Forest Ecology and Management 251: 205–216. doi:10.1016/j.foreco.2007.06.005. Retrieved 28 February 2012.

3

Air Pollutant: A Quantitative Study

The major components of air pollution are discussed in this chapter. It deals with air pollutants, which are substances in the air that can have adverse effects on humans as well as the ecosystem. Pollutants can be natural or man-made, and are classified as primary or secondary. This chapter encompasses and incorporates the major components and key concepts of air pollution, and provides a complete understanding.

Pollutant

Surface runoff, also called nonpoint source pollution, from a farm field in Iowa, United States during a rain storm. Topsoil as well as farm fertilizers and other potential pollutants run off unprotected farm fields when heavy rains occur.

A pollutant is a substance or energy introduced into the environment that has undesired effects, or adversely affects the usefulness of a resource. A pollutant may cause long- or short-term damage by changing the growth rate of plant or animal species, or by interfering with human amenities, comfort, health, or property values. Some pollutants are biodegradable and therefore will not persist in the environment in the long term. However the degradation products of some pollutants are themselves polluting such as the products DDE and DDD produced from degradation of DDT.

Different Types of Pollutants in Nature

Stock Pollutants

Pollutants, towards which the environment has little or no absorptive capacity are called *stock pollutants*. (e.g. persistent synthetic chemicals, non-biodegradable plastics, and heavy metals). Stock pollutants accumulate in the environment over time. The damage they cause increases as more pollutant is emitted, and persists as the pollutant accumulates. Stock pollutants can create a burden for future generations, by passing on damage that persists well after the 'benefits' received from incurring that damage; have been forgotten.

Fund Pollutants

Fund pollutants are those for which the environment has some absorptive capacity. Fund pollutants do not cause damage to the environment unless the emission rate exceeds the receiving environment's absorptive capacity (e.g. carbon dioxide, which is absorbed by plants and oceans). Fund pollutants are not destroyed, but rather converted into less harmful substances, or diluted/ dispersed to non-harmful concentrations.

Notable Pollutants

Notable pollutants include the following groups:

- Heavy metals
- Persistent organic pollutants POP
- Environmental Persistent Pharmaceutical Pollutants EPPP
- Polycyclic aromatic hydrocarbons
- Volatile organic compounds
- Environmental xenobiotics

Light Pollution

Light pollution is the impact that anthropogenic light has on the visibility of the night sky. It also encompasses

The night sky viewed from Luhasoo bog, Estonia with light pollution in background.

ecological light pollution which describes the effect of artificial light on individual organisms and on the structure of ecosystems as a whole.

Zones of Influence

Pollutants can also be defined by their zones of influence, both horizontally and vertically.

Horizontal Zone

The horizontal zone refers to the area that is damaged by a pollutant. Local pollutants cause damage near the emission source. Regional pollutants cause damage further from the emission source.

Vertical Zone

The vertical zone is referred to whether the damage is ground-level or atmospheric. Surface pollutants cause damage by concentrations of the pollutant accumulating near the Earth's surface Global pollutants cause damage by concentrations in the atmosphere.

Regulation

International

Pollutants can cross international borders and therefore international regulations are needed for their control. The Stockholm Convention on Persistent Organic Pollutants, which entered into force in 2004, is an international legally binding agreement for the control of persistent organic pollutants. Pollutant Release and Transfer Registers (PRTR) are systems to collect and disseminate information on environmental releases and transfers of toxic chemicals from industrial and other facilities.

European Union

The European Pollutant Emission Register is a type of PRTR providing access to information on the annual emissions of industrial facilities in the Member States of the European Union, as well as Norway.

United States

Clean Air Act standards. Under the Clean Air Act, the National Ambient Air Quality Standards (NAAQS) are developed by the Environmental Protection Agency (EPA) for six common air pollutants, also called "criteria pollutants": particulates; smog and ground-level ozone; carbon monoxide; sulfur oxides; nitrogen oxides; and lead. The National Emissions Standards for Hazardous Air Pollutants are additional emission standards that are set by EPA for toxic air pollutants.

Clean Water Act standards. Under the Clean Water Act, EPA promulgated national standards for municipal sewage treatment plants, also called *publicly owned treatment works,* in the *Secondary Treatment Regulation.* National standards for industrial dischargers are called *Effluent guidelines* (for existing sources) and New Source Performance Standards, and currently cover over 50 industrial categories. In addition, the Act requires states to publish water quality standards for individual water bodies to provide additional protection where the national standards are insufficient.

RCRA standards. The Resource Conservation and Recovery Act (RCRA) regulates the management, transport and disposal of municipal solid waste, hazardous waste and underground storage tanks.

A tank or piping network that has at least 10 percent of its volume underground is known as an underground storage tank (UST). They often store substances such as petroleum, that are harmful to the surrounding environment should it become contaminated.

Air Pollutant Concentrations

Air pollutant concentrations, as measured or as calculated by air pollution dispersion modeling, must often be converted or corrected to be expressed as required by the regulations issued by various governmental agencies. Regulations that define and limit the concentration of pollutants in the ambient air or in gaseous emissions to the ambient air are issued by various national and state (or provincial) environmental protection and occupational health and safety agencies.

Such regulations involve a number of different expressions of concentration. Some express the concentrations as ppmv (parts per million by volume) and some express the concentrations as mg/m^3 (milligrams per cubic meter), while others require adjusting or correcting the concentrations to reference conditions of moisture content, oxygen content or carbon dioxide content. This article presents methods for converting concentrations from ppmv to mg/m^3 (and vice versa) and for correcting the concentrations to the required reference conditions.

All of the concentrations and concentration corrections in this article apply only to air and other gases. They are not applicable for liquids.

Carbon dioxide in Earth's atmosphere if *half* of global-warming emissions are *not* absorbed. (NASA simulation; 9 November 2015)

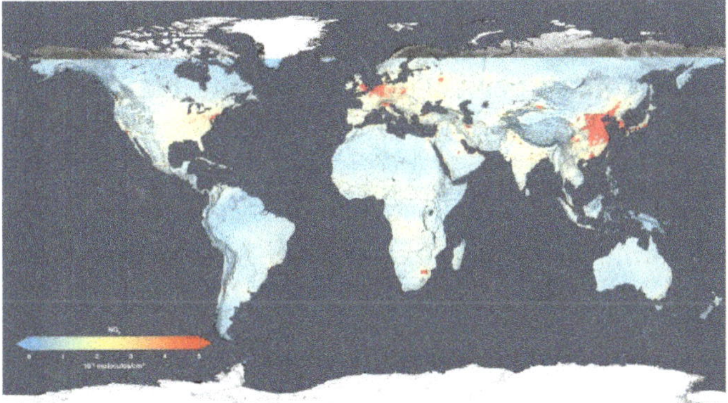

Nitrogen dioxide 2014 - global air quality levels (released 14 December 2015).

Converting Air Pollutant Concentrations

The conversion equations depend on the temperature at which the conversion is wanted (usually about 20 to 25 °C). At an ambient sea level atmospheric pressure of 1 atm (101.325 kPa or 1.01325 bar), the general equation is:

$$\text{ppmv} = \text{mg}/\text{m}^3 \cdot \frac{(0.08205 \cdot T)}{M}$$

and for the reverse conversion:

$$\text{mg}/\text{m}^3 = \text{ppmv} \cdot \frac{M}{(0.08205 \cdot T)}$$

where:

mg/m^3= milligrams of pollutant per cubic meter of air at sea level atmospheric pressure and T

ppmv= air pollutant concentration, in parts per million by volume

T= ambient temperature in K = 273. + °C

0.08205= Universal gas constant in atm·m^3/(kmol·K)

M= molecular mass (or molecular weight) of the air pollutant

Notes:

1 atm = absolute pressure of 101.325 kPa or 1.01325 bar

- mol = gram mole and kmol = 1000 gram moles
- Pollution regulations in the United States typically reference their pollutant limits to an ambient temperature of 20 to 25 °C as noted above. In most other nations, the reference ambient temperature for pollutant limits may be 0 °C or other values.
- Although ppmv and mg/m^3 have been used for the examples in all of the following sections, concentrations such as ppbv (i.e., parts per billion by volume), volume percent, mole percent and many others may also be used for gaseous pollutants.
- Particulate matter (PM) in the atmospheric air or in any other gas cannot be expressed in terms of ppmv, ppbv, volume percent or mole percent. PM is most usually (but not always) expressed as mg/m^3 of air or other gas at a specified temperature and pressure.
- For gases, volume percent = mole percent
- 1 volume percent = 10,000 ppmv (i.e., parts per million by volume) with a million being defined as 10^6.
- Care must be taken with the concentrations expressed as ppbv to differentiate between the British billion which is 10^{12} and the USA billion which is 10^9 (also referred to as the long scale and short scale billion, respectively).

Correcting Concentrations for Altitude

Air pollutant concentrations expressed as mass per unit volume of atmospheric air (e.g., mg/m^3, $\mu g/m^3$, etc.) at sea level will decrease with increasing altitude. The concentration decrease is directly proportional to the pressure decrease with increasing altitude. Some governmental regulatory jurisdictions require industrial sources of air pollution to comply with sea level standards corrected for altitude. In other words, industrial air pollution sources located at altitudes well above sea level must comply with significantly more stringent air quality standards than sources located at sea level (since it is more difficult to comply with lower standards). For example, New Mexico's Department of the Environment has a regulation with such a requirement.

The change of atmospheric pressure with altitude (<20 km) can be obtained from this equation:

$$P_h = P \cdot \left(\frac{288 - 6.5h}{288}\right)^{5.2558}$$

Given an air pollutant concentration at sea-level atmospheric pressure, the concentration at higher altitudes can be obtained from this equation:

$$C_h = C \cdot \left(\frac{288 - 6.5h}{288}\right)^{5.2558}$$

where:

h= altitude, in km

P= atmospheric pressure at sea level

P_h= atmospheric pressure at altitude h

C= Air pollutant concentration, in mass per unit volume at sea level atmospheric pressure and specified temperature T

C_h= Concentration, in mass per unit volume at altitude h and specified temperature T

As an example, given an air pollutant concentration of 260 mg/m^3 at sea level, calculate the equivalent pollutant concentration at an altitude of 2800 meters:

$$C_h = 260 \times [\{288 - (6.5)(2.8)\} / 288]^{5.2558} = 260 \times 0.71 = 185 \text{ mg/m}^3$$

Note:

- The above equation for the decrease of air pollution concentrations with increasing altitude is applicable only for about the first 10 km of altitude in the troposphere (the lowest atmospheric layer) and is estimated to have a maximum error of about 3 percent. However, 10 km of altitude is sufficient for most purposes involving air pollutant concentrations.

Correcting Concentrations for Reference Conditions

Many environmental protection agencies have issued regulations that limit the concentration of pollutants in gaseous emissions and define the reference conditions applicable to those concentration limits. For example, such a regulation might limit the concentration of NOx to 55 ppmv in a dry combustion exhaust gas (at a specified reference temperature and pressure) corrected to 3 volume percent O_2 in the dry gas. As another example, a regulation might limit the concentration of total particulate matter to 200 mg/m³ of an emitted gas (at a specified reference temperature and pressure) corrected to a dry basis and further corrected to 12 volume percent CO_2 in the dry gas.

Environmental agencies in the USA often use the terms "dscf" or "scfd" to denote a "standard" cubic foot of dry gas. Likewise, they often use the terms "dscm" or "scmd" to denote a "standard" cubic meter of gas. Since there is no universally accepted set of "standard" temperature and pressure, such usage can be and is very confusing. It is strongly recommended that the reference temperature and pressure always be clearly specified when stating gas volumes or gas flow rates.

Correcting to a Dry Basis

If a gaseous emission sample is analyzed and found to contain water vapor and a pollutant concentration of say 40 ppmv, then 40 ppmv should be designated as the "wet basis" pollutant concentration. The following equation can be used to correct the measured "wet basis" concentration to a "dry basis" concentration:

$$C_{\text{dry basis}} = \frac{C_{\text{wet basis}}}{1 - w}$$

where:

C = Concentration of the air pollutant in the emitted gas

w = fraction of the emitted exhaust gas, by volume, which is water vapor

As an example, a wet basis concentration of 40 ppmv in a gas having 10 volume percent water vapor would have a:

$$C_{\text{dry basis}} = 40 \div (1 - 0.10) = 44.4 \text{ ppmv.}$$

Correcting to a Reference Oxygen Content

The following equation can be used to correct a measured pollutant concentration in a dry emitted gas with a measured O_2 content to an equivalent pollutant concentration in a dry emitted gas with a specified reference amount of O_2:

$$C_r = C_m \cdot \frac{(20.9 - \text{reference volume}\%O_2)}{(20.9 - \text{measured volume}\%O_2)}$$

where:

C_r = corrected concentration of a dry gas with a specified reference volume % O_2

C_m = measured concentration in a dry gas having a measured volume % O_2

As an example, a measured NOx concentration of 45 ppmv in a dry gas having 5 volume % O_2 is:

$$45 \times (20.9 - 3) \div (20.9 - 5) = 50.7 \text{ ppmv of NOx}$$

when corrected to a dry gas having a specified reference O_2 content of 3 volume %.

Note:

- The measured gas concentration C_m must first be corrected to a dry basis before using the above equation.

Correcting to a Reference Carbon Dioxide Content

The following equation can be used to correct a measured pollutant concentration in an emitted gas (containing a measured CO_2 content) to an equivalent pollutant concentration in an emitted gas containing a specified reference amount of CO_2

$$C_r = C_m \cdot \frac{(\text{reference volume}\%CO_2)}{(\text{measured volume}\%CO_2)}$$

where:

C_r= corrected concentration of a dry gas having a specified reference volume % CO_2

C_m= measured concentration of a dry gas having a measured volume % CO_2

As an example, a measured particulates concentration of 200 mg/m^3 in a dry gas that has a measured 8 volume % CO_2 is:

$$200 \times (12 \div 8) = 300 \text{ mg/m}^3$$

when corrected to a dry gas having a specified reference CO_2 content of 12 volume %.

Tropospheric Ozone

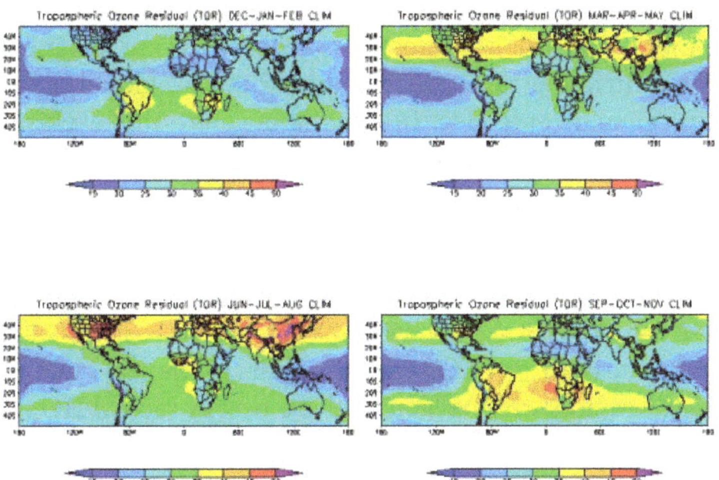

Seasonal average vertical columns of tropospheric ozone in Dobson units over the period 1979 to 2000. In June to August photochemical ozone production causes very high concentrations over the East Coast of the USA and China.

Ozone (O_3) is a constituent of the troposphere (it is also an important constituent of some regions of the stratosphere commonly known as the ozone layer). The troposphere extends from the Earth's surface to between 12 and 20 kilometers above sea level and consists of many layers. Ozone is more concentrated above the mixing layer, or ground layer. Ground-level ozone, though less concentrated than ozone aloft, is more of a problem because of its health effects.

Photochemical and chemical reactions involving it drive many of the chemical processes that occur in the atmosphere by day and by night. At abnormally high concentrations brought about by human activities (largely incomplete combustion of fossil fuels, such as gasoline, diesel, etc.), it is a pollutant, and a constituent of smog. Many highly energetic reactions produce it, ranging from combustion to photocopying. Often laser printers will have a smell of ozone, which in high concentrations is toxic. Ozone is a powerful oxidizing agent readily reacting with other chemical compounds to make many possibly toxic oxides.

Tropospheric ozone is a greenhouse gas and initiates the chemical removal of methane and other hydrocarbons from the atmosphere. Thus, its concentration affects how long these compounds remain in the air.

Measurement

Satellites are able to measure tropospheric ozone. Measurements specifically of ground-level ozone require *in situ* monitoring technology.

Formation

The majority of tropospheric ozone formation occurs when nitrogen oxides (NOx), carbon monoxide (CO) and volatile organic compounds (VOCs), react in the atmosphere in the presence of sunlight. NOx, CO, and VOCs are called ozone precursors. Motor vehicle exhaust, industrial emissions, and chemical solvents are the major anthropogenic sources of these chemicals. Another source is windshield washer fluid. Although these precursors often originate in urban areas, winds can carry NOx hundreds of kilometers, causing ozone formation to occur in less populated regions as well. Methane, a VOC whose atmospheric concentration has increased tremendously during the last century, contributes to ozone formation but on a global scale rather than in local or regional photochemical smog episodes. In situations where this exclusion of methane from the VOC group of substances is not obvious, the term Non-Methane VOC (NMVOC) is often used.

The chemical reactions involved in tropospheric ozone formation are a series of complex cycles in which carbon monoxide and VOCs are oxidised to water vapour and carbon dioxide. The reactions involved in this process are illustrated here with CO but similar reactions occur for VOC as well. The oxidation begins with the reaction of CO with the hydroxyl radical (·OH). The radical intermediate formed by this reacts rapidly with oxygen to give a peroxy radical $HO_2^{\cdot}$

$$OH + CO \rightarrow {}^{\cdot}HOCO$$

$$HOCO + O_2 \rightarrow HO_2^{\cdot} + CO_2$$

Peroxy radicals then go on to react with NO to give NO_2 which is photolysed to give atomic oxygen and through reaction with oxygen a molecule of ozone:

$$HO_2^{\bullet} + NO \rightarrow {}^{\bullet}OH + NO_2$$

$$NO_2 + h\nu \rightarrow NO + O(^3P)$$

$$O(^3P) + O_2 \rightarrow O_3$$

The balance of this sequence of chemical reactions is:

$$CO + 2O_2 + h\nu \rightarrow CO_2 + O_3$$

The amount of ozone produced through these reactions can be calculated using the Leighton relationship.

This cycle involving HO_x and NO_x is terminated by the reaction of OH with NO_2 to form nitric acid or by the reaction of peroxy radicals with each other to form peroxides. The chemistry involving VOCs is much more complex but the same reaction of peroxy radicals oxidizing NO to NO_2 is the critical step leading to ozone formation.

Health Effects

Health effects depend on ozone precursors, which is a group of pollutants, primarily generated during the combustion of fossil fuels. Reaction with daylight ultraviolet (UV) rays and these precursors create ground-level ozone pollution (Tropospheric Ozone). Ozone is known to have the following health effects at concentrations common in urban air:

- Irritation of the respiratory system, causing coughing, throat irritation, and/or an uncomfortable sensation in the chest.
- Reduced lung function, making it more difficult to breathe deeply and vigorously. Breathing may become more rapid and more shallow than normal, and a person's ability to engage in vigorous activities may be limited.
- Aggravation of asthma. When ozone levels are high, more people with asthma have attacks that require a doctor's attention or use of medication. One reason this happens is that ozone makes people more sensitive to allergens, which in turn trigger asthma attacks.
- Increased susceptibility to respiratory infections.
- Inflammation and damage to the lining of the lungs. Within a few days, the damaged cells are shed and replaced much like the skin peels after a sunburn. Animal studies suggest that if this type of inflammation happens repeatedly over a long time period (months, years, a lifetime), lung tissue may become permanently scarred, resulting in permanent loss of lung function and a lower quality of life.

A statistical study of 95 large urban communities in the United States found significant association between ozone levels and premature death. The study estimated that a one-third reduction in urban ozone concentrations would save roughly 4000 lives per year (Bell et al., 2004). Tropospheric Ozone causes approximately 22,000 premature deaths per year in 25 countries in the European Union. (WHO, 2008)

Problem Areas

The United States Environmental Protection Agency has developed an Air Quality index to help explain air pollution levels to the general public. 8-hour average ozone mole fractions of 76 to 95 nmol/mol are described as "Unhealthy for Sensitive Groups", 96 nmol/mol to 115 nmol/mol as "unhealthy" and 116 nmol/mol to 404 nmol/mol as "very unhealthy" . The EPA has designated over 300 counties of the United States, clustered around the most heavily populated areas (especially in California and the Northeast), as failing to comply with the National Ambient Air Quality Standards.

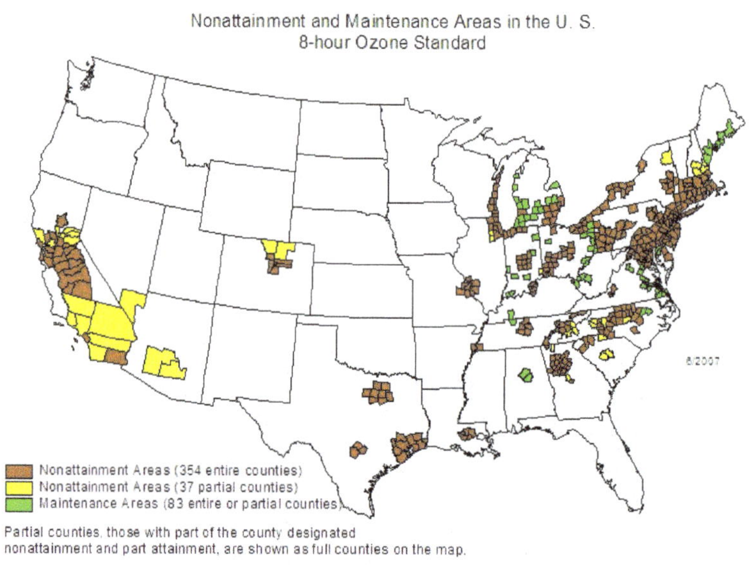

Climate Change

Melting of sea ice releases molecular chlorine, which reacts with UV radiation to produce chlorine radicals. Because chlorine radicals are highly reactive, they can expedite the degradation of methane and tropospheric ozone and the oxidation of mercury to more toxic forms. Ozone production rises during heat waves, because plants absorb less ozone. It is estimated that curtailed ozone absorption by plants is responsible for the loss of 460 lives in the UK in the hot summer of 2006. A similar investigation to assess the joint effects of ozone and heat during the European heat waves in 2003, concluded that these appear to be additive.

References

- David A. Lewandowski (1999). Design of Thermal Oxidation Systems for Volatile Organic Compounds (1st ed.). CRC Press. ISBN 1-56670-410-3.
- Amann, Derwent, Forsberg et al. (2008): Health risks of ozone from long-range transboundary air pollution. Publication of World Health Organization Europe. ISBN 978-92-890-4289-5
- Seinfeld, John H.; Pandis, Spyros N (1998). Atmospheric Chemistry and Physics - From Air Pollution to Climate Change. John Wiley and Sons, Inc. ISBN 0-471-17816-0
- St. Fleur, Nicholas (10 November 2015). "Atmospheric Greenhouse Gas Levels Hit Record, Report Says". New York Times. Retrieved 11 November 2015.

- Ritter, Karl (9 November 2015). “UK: In 1st, global temps average could be 1 degree C higher”. AP News. Retrieved 11 November 2015.
- Cole, Steve; Gray, Ellen (14 December 2015). “New NASA Satellite Maps Show Human Fingerprint on Global Air Quality”. NASA. Retrieved 14 December 2015.
- Jin Liao; et al. (January 2014). “High levels of molecular chlorine in the Arctic atmosphere”. Nature Geoscience Letter. doi:10.1038/ngeo2046. Retrieved January 14, 2014.

4

Effects of Air Pollution on Ecology and Human Health

The effects of air pollution can be very harsh, both on the human health as well as ecology. Air pollution is a concern equally for an individual and an entire population. It is a significant risk factor for a number of pollution-related diseases and health conditions including respiratory infections, heart disease, stroke and lung cancer.

Indoor Air Quality

Indoor air quality (IAQ) is a term which refers to the air quality within and around buildings and structures, especially as it relates to the health and comfort of building occupants. IAQ can be affected by gases (including carbon monoxide, radon, volatile organic compounds), particulates, microbial contaminants (mold, bacteria), or any mass or energy stressor that can induce adverse health conditions. Source control, filtration and the use of ventilation to dilute contaminants are the primary methods for improving indoor air quality in most buildings. Residential units can further improve indoor air quality by routine cleaning of carpets and area rugs.

A common air filter, being cleaned with a vacuum cleaner

Determination of IAQ involves the collection of air samples, monitoring human exposure to pollutants, collection of samples on building surfaces, and computer modelling of air flow inside buildings.

IAQ is part of indoor environmental quality (IEQ), which includes IAQ as well as other physical and psychological aspects of life indoors (e.g., lighting, visual quality, acoustics, and thermal comfort).

Indoor air pollution in developing nations is a major health hazard. A major source of indoor air pollution in developing countries is the burning of biomass (e.g. wood, charcoal, dung, or crop residue) for heating and cooking. The resulting exposure to high levels of particulate matter resulted in between 1.5 million and 2 million deaths in 2000.

Common Pollutants

Second-hand Smoke

Second-hand smoke is tobacco smoke which affects other people other than the 'active' smoker. Second-hand tobacco smoke includes both a gaseous and a particulate phase, with particular hazards arising from levels of carbon monoxide (as indicated below) and very small particulates (at PM2.5 size) which get past the lung's natural defenses. The only certain method to improve indoor air quality as regards second-hand smoke is the implementation of comprehensive smoke-free laws.

Radon

Radon is an invisible, radioactive atomic gas that results from the radioactive decay of radium, which may be found in rock formations beneath buildings or in certain building materials themselves. Radon is probably the most pervasive serious hazard for indoor air in the United States and Europe, probably responsible for tens of thousands of deaths from lung cancer each year. There are relatively simple test kits for do-it-yourself radon gas testing, but if a home is for sale the testing must be done by licensed person in some U.S. states. Radon gas enters buildings as a soil gas and is a heavy gas and thus will tend to accumulate at the lowest level. Radon may also be introduced into a building through drinking water particularly from bathroom showers. Building materials can be a rare source of radon, but little testing is carried out for stone, rock or tile products brought into building sites; radon accumulation is greatest for well insulated homes. The half life for radon is 3.8 days, indicating that once the source is removed, the hazard will be greatly reduced within a few weeks. Radon mitigation methods include sealing concrete slab floors, basement foundations, water drainage systems, or by increasing ventilation. They are usually cost effective and can greatly reduce or even eliminate the contamination and the associated health risks.

Molds and Other Allergens

These biological chemicals can arise from a host of means, but there are two common classes: (a) moisture induced growth of mold colonies and (b) natural substances released into the air such as animal dander and plant pollen. Mold is always associated with moisture, and its growth can be inhibited by keeping humidity levels below 50%. Moisture buildup inside buildings may arise from water penetrating compromised areas of the building envelope or skin, from plumbing leaks, from condensation due to improper ventilation, or from ground moisture penetrating a building part. In areas where cellulosic materials (paper and wood, including drywall) become moist and fail to dry within 48 hours, mold mildew can propagate and release allergenic spores into the air.

In many cases, if materials have failed to dry out several days after the suspected water event, mold growth is suspected within wall cavities even if it is not immediately visible. Through a mold investigation, which may include destructive inspection, one should be able to determine the presence or absence of mold. In a situation where there is visible mold and the indoor air quality may have been compromised, mold remediation may be needed. Mold testing and inspections should be carried out by an independent investigator to avoid any conflict of interest and to insure accurate results; free mold testing offered by remediation companies is not recommended.

There are some varieties of mold that contain toxic compounds (mycotoxins). However, exposure to hazardous levels of mycotoxin via inhalation is not possible in most cases, as toxins are produced by the fungal body and are not at significant levels in the released spores. The primary hazard of mold growth, as it relates to indoor air quality, comes from the allergenic properties of the spore cell wall. More serious than most allergenic properties is the ability of mold to trigger episodes in persons that already have asthma, a serious respiratory disease.

Carbon Monoxide

One of the most acutely toxic indoor air contaminants is carbon monoxide (CO), a colourless, odourless gas that is a byproduct of incomplete combustion of fossil fuels. Common sources of carbon monoxide are tobacco smoke, space heaters using fossil fuels, defective central heating furnaces and automobile exhaust. By depriving the brain of oxygen, high levels of carbon monoxide can lead to nausea, unconsciousness and death. According to the American Conference of Governmental Industrial Hygienists (ACGIH), the time-weighted average (TWA) limit for carbon monoxide (630-08-0) is 25 ppm.

Indoor levels of CO are systematically improving due to increasing implementation of smoke-free laws.

Volatile Organic Compounds

Volatile organic compounds (VOCs) are emitted as gases from certain solids or liquids. VOCs include a variety of chemicals, some of which may have short- and long-term adverse health effects. Concentrations of many VOCs are consistently higher indoors (up to ten times higher) than outdoors. VOCs are emitted by a wide array of products numbering in the thousands. Examples include: paints and lacquers, paint strippers, cleaning supplies, pesticides, building materials and furnishings, office equipment such as copiers and printers, correction fluids and carbonless copy paper, graphics and craft materials including glues and adhesives, permanent markers, and photographic solutions.

Chlorinated drinking water releases chloroform when hot water is used in the home. Benzene is emitted from fuel stored in attached garages. Overheated cooking oils emit acrolein and formaldehyde. A meta-analysis of 77 surveys of VOCs in homes in the US found the top ten riskiest indoor air VOCs were acrolein, formaldehyde, benzene, hexachlorobutadiene, acetaldehyde, 1,3-butadiene, benzyl chloride, 1,4-dichlorobenzene, carbon tetrachloride, acrylonitrile, and vinyl chloride. These compounds exceeded health standards in most homes.

Organic chemicals are widely used as ingredients in household products. Paints, varnishes, and wax all contain organic solvents, as do many cleaning, disinfecting, cosmetic, degreasing, and hobby products. Fuels are made up of organic chemicals. All of these products can release organic compounds during usage, and, to some degree, when they are stored. Testing emissions from building materials used indoors has become increasingly common for floor coverings, paints, and many other important indoor building materials and finishes.

Several initiatives envisage to reduce indoor air contamination by limiting VOC emissions from products. There are regulations in France and in Germany, and numerous voluntary ecolabels and rating systems containing low VOC emissions criteria such as EMICODE, M1, Blue Angel and Indoor Air Comfort in Europe, as well as California Standard CDPH Section 01350 and several others in the USA. These initiatives changed the marketplace where an increasing number of low-emitting products has become available during the last decades.

At least 18 Microbial VOCs (MVOCs) have been characterised including 1-octen-3-ol, 3-methylfuran, 2-pentanol, 2-hexanone, 2-heptanone, 3-octanone, 3-octanol, 2-octen-1-ol, 1-octene, 2-pentanone, 2-nonanone, borneol, geosmin, 1-butanol, 3-methyl-1-butanol, 3-methyl-2-butanol, and thujopsene. The first of these compounds is called mushroom alcohol. The last four are products of *Stachybotrys chartarum*, which has been linked with sick building syndrome.

Legionella

Legionellosis or Legionnaire's Disease is caused by a waterborne bacterium *Legionella* that grows best in slow-moving or still, warm water. The primary route of exposure is through the creation of an aerosol effect, most commonly from evaporative cooling towers or showerheads. A common source of Legionella in commercial buildings is from poorly placed or maintained evaporative cooling towers, which often release water in an aerosol which may enter nearby ventilation intakes. Outbreaks in medical facilities and nursing homes, where patients are immuno-suppressed and immuno-weak, are the most commonly reported cases of Legionellosis. More than one case has involved outdoor fountains in public attractions. The presence of Legionella in commercial building water supplies is highly under-reported, as healthy people require heavy exposure to acquire infection.

Legionella testing typically involves collecting water samples and surface swabs from evaporative cooling basins, shower heads, faucets/taps, and other locations where warm water collects. The samples are then cultured and colony forming units (cfu) of Legionella are quantified as cfu/Liter.

Legionella is a parasite of protozoans such as amoeba, and thus requires conditions suitable for both organisms. The bacterium forms a biofilm which is resistant to chemical and antimicrobial treatments, including chlorine. Remediation for Legionella outbreaks in commercial buildings vary, but often include very hot water flushes (160 °F; 70 °C), sterilisation of standing water in evaporative cooling basins, replacement of shower heads, and in some cases flushes of heavy metal salts. Preventative measures include adjusting normal hot water levels to allow for 120 °F at the tap, evaluating facility design layout, removing faucet aerators, and periodic testing in suspect areas.

Other Bacteria

There are many bacteria of health significance found in indoor air and on indoor surfaces. The role of microbes in the indoor environment is increasingly studied using modern gene-based analysis of environmental samples. Currently efforts are under way to link microbial ecologists and indoor air scientists to forge new methods for analysis and to better interpret the results.

"There are approximately ten times as many bacterial cells in the human flora as there are human cells in the body, with large numbers of bacteria on the skin and as gut flora." A large fraction of the bacteria found in indoor air and dust are shed from humans. Among the most important bacteria known to occur in indoor air are Mycobacterium tuberculosis, Staphylococcus aureus, Streptococcus pneumoniae.

Asbestos Fibers

Many common building materials used before 1975 contain asbestos, such as some floor tiles, ceiling tiles, shingles, fireproofing, heating systems, pipe wrap, taping muds, mastics, and other insulation materials. Normally, significant releases of asbestos fiber do not occur unless the building materials are disturbed, such as by cutting, sanding, drilling, or building remodelling. Removal of asbestos-containing materials is not always optimal because the fibers can be spread into the air during the removal process. A management program for intact asbestos-containing materials is often recommended instead.

When asbestos-containing material is damaged or disintegrates, microscopic fibers are dispersed into the air. Inhalation of asbestos fibers over long exposure times is associated with increased incidence of lung cancer, in particular the specific form mesothelioma. The risk of lung cancer from inhaling asbestos fibers is also greater to smokers. The symptoms of the disease do not usually appear until about 20 to 30 years after the first exposure to asbestos.

Asbestos is found in older homes and buildings, but occurs most commonly in schools and industrial settings. The US Federal Government (www.osha.gov) and some states have set standards for acceptable levels of asbestos fibers in indoor air. There are particularly stringent regulations applicable to schools.

Carbon Dioxide

Carbon dioxide (CO_2) is a relatively easy to measure surrogate for indoor pollutants emitted by humans, and correlates with human metabolic activity. Carbon dioxide at levels that are unusually high indoors may cause occupants to grow drowsy, to get headaches, or to function at lower activity levels. Humans are the main indoor source of carbon dioxide in most buildings. Indoor CO_2 levels are an indicator of the adequacy of outdoor air ventilation relative to indoor occupant density and metabolic activity.

To eliminate most complaints, the total indoor CO_2 level should be reduced to a difference of less than 600 ppm above outdoor levels. The National Institute for Occupational Safety and Health (NIOSH) considers that indoor air concentrations of carbon dioxide that exceed 1,000 ppm are a marker suggesting inadequate ventilation. The UK standards for schools say that carbon dioxide in all teaching and learning spaces, when measured at seated head height and averaged over the whole day should not exceed 1,500 ppm. The whole day refers to normal school hours (i.e. 9:00am to 3:30pm) and includes unoccupied periods such as lunch breaks. In Hong Kong, the EPD established indoor air quality objectives for office buildings and public places in which a carbon dioxide level below 1,000 ppm is considered to be good. European standards limit carbon dioxide to 3,500 ppm. OSHA limits carbon dioxide concentration in the workplace to 5,000 ppm for prolonged periods, and 35,000 ppm for 15 minutes. These higher limits are concerned with avoiding loss of consciousness (fainting), and do not address impaired cognitive performance and energy, which begin to occur at lower concentrations of carbon dioxide.

Carbon dioxide concentrations increase as a result of human occupancy, but lag in time behind cumulative occupancy and intake of fresh air. The lower the air exchange rate, the slower the buildup of carbon dioxide to quasi "steady state" concentrations on which the NIOSH and UK guidance are based. Therefore, measurements of carbon dioxide for purposes of assessing the adequacy of ventilation need to be made after an extended period of steady occupancy and ventilation - in schools at least 2 hours, and in offices at least 3 hours - for concentrations to be a reasonable indicator of ventilation adequacy. Portable instruments used to measure carbon dioxide should be calibrated frequently, and outdoor measurements used for calculations should be made close in time to indoor measurements. Corrections for temperature effects on measurements made outdoors may also be necessary.

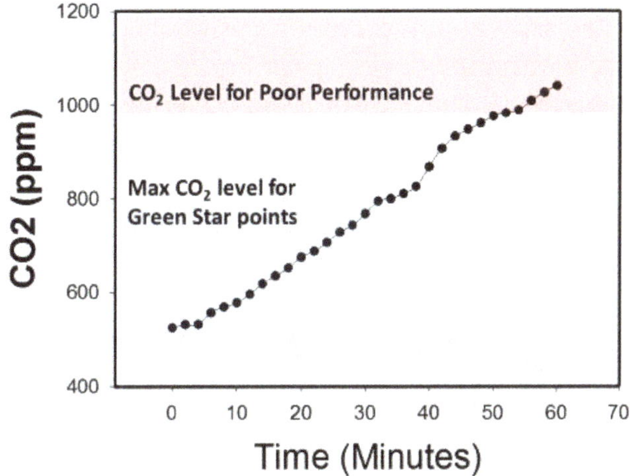

CO2 levels in an enclosed office room can increase to over 1,000 ppm within 45 minutes.

Carbon dioxide concentrations in closed or confined rooms can increase to 1,000 ppm within 45 minutes of enclosure. For example, in a 3.5-by-4-metre (11 ft × 13 ft) sized office, atmospheric carbon dioxide increased from 500 ppm to over 1,000 ppm within 45 minutes of ventilation cessation and closure of windows and doors.

Ozone

Ozone is produced by ultraviolet light from the Sun hitting the Earth's atmosphere (especially in the ozone layer), lightning, certain high-voltage electric devices (such as air ionizers), and as a by-product of other types of pollution.

Ozone exists in greater concentrations at altitudes commonly flown by passenger jets. Reactions between ozone and onboard substances, including skin oils and cosmetics, can produce toxic chemicals as by-products. Ozone itself is also irritating to lung tissue and harmful to human health. Larger jets have ozone filters to reduce the cabin concentration to safer and more comfortable levels.

Outdoor air used for ventilation may have sufficient ozone to react with common indoor pollutants as well as skin oils and other common indoor air chemicals or surfaces. Particular concern is warranted when using "green" cleaning products based on citrus or terpene extracts, because these chemicals react very quickly with ozone to form toxic and irritating chemicals as well as fine and ultrafine particles. Ventilation with outdoor air containing elevated ozone concentrations may complicate remediation attempts.

Prompt Cognitive Deficits

In 2015, experimental studies reported the detection of significant episodic (situational) cognitive impairment from impurities in the air breathed by test subjects who were not informed about changes in the air quality. Researchers at the Harvard University and SUNY Upstate Medical University and Syracuse University measured the cognitive performance of 24 participants in three different controlled laboratory atmospheres that simulated those found in "conventional" and "green" buildings, as well as green buildings with enhanced ventilation. Performance was evaluated objectively using the widely used Strategic Management Simulation software simulation tool, which is a well-validated assessment test for executive decision-making in an unconstrained situation allowing initiative and improvisation. Significant deficits were observed in the performance scores achieved in increasing concentrations of either volatile organic compounds (VOCs) or carbon dioxide, while keeping other factors constant. The highest impurity levels reached are not uncommon in some classroom or office environments.

Effect of Indoor Plants

Spider plants *(Chlorophytum comosum)* absorb some airborne contaminants

Houseplants together with the medium in which they are grown can reduce components of indoor air pollution, particularly volatile organic compounds (VOC) such as benzene, toluene, and xylene. Plants remove CO_2 and release oxygen and water, although the quantitative impact for house plants is small. Most of the effect is attributed to the growing medium alone, but even this effect has finite limits associated with the type and quantity of medium and the flow of air through the medium. The effect of house plants on VOC concentrations was investigated in one study, done in a static chamber, by NASA for possible use in space colonies. The results showed that the removal of the challenge chemicals was roughly equivalent to that provided by the ventilation that occurred in a very energy efficient dwelling with a very low ventilation rate, an air exchange rate of about 1/10 per hour. Therefore, air leakage in most homes, and in non-residential buildings too, will generally remove the chemicals faster than the researchers reported for the plants tested by NASA. The most effective household plants reportedly included aloe vera, English ivy, and Boston fern for removing chemicals and biological compounds.

Plants also appear to reduce airborne microbes, molds, and increase humidity. However, the increased humidity can itself lead to increased levels of mold and even VOCs.

When CO2 concentrations are elevated indoors relative to outdoor concentrations, it is only an indicator that ventilation is inadequate to remove metabolic products associated with human occupancy. Plants require CO2 to grow and release oxygen when they consume CO2. A study published in the journal *Environmental Science & Technology* considered uptake rates of ketones and aldehydes by the peace lily (Spathiphyllum clevelandii) and golden pothos (Epipremnum aureum.) Akira Tani and C. Nicholas Hewitt found "Longer-term fumigation results revealed that the total uptake amounts were 30–100 times as much as the amounts dissolved in the leaf, suggesting that volatile organic carbons are metabolized in the leaf and/or translocated through the petiole." It is worth noting the researchers sealed the plants in Teflon bags. "No VOC loss was detected from the bag when the plants were absent. However, when the plants were in the bag, the levels of aldehydes and ketones both decreased slowly but continuously, indicating removal by the plants". Studies done in sealed bags do not faithfully reproduce the conditions in the indoor environments of interest. Dynamic conditions with outdoor air ventilation and the processes related to the surfaces of the building itself and its contents as well as the occupants need to be studied.

While results do indicate house plants may be effective at removing some VOCs from air supplies, a review of studies between 1989 and 2006 on the performance of houseplants as air cleaners, presented at the Healthy Buildings 2009 conference in Syracuse, NY, concluded "...indoor plants have little, if any, benefit for removing indoor air of VOC in residential and commercial buildings."

Since high humidity is associated with increased mold growth, allergic responses, and respiratory responses, the presence of additional moisture from houseplants may not be desirable in all indoor settings.

HVAC Design

Environmentally sustainable design concepts also include aspects related to the commercial and residential heating, ventilation and air-conditioning (HVAC) industry. Among several considerations, one of the topics attended to is the issue of indoor air quality throughout the design and construction stages of a building's life.

One technique to reduce energy consumption while maintaining adequate air quality, is demand controlled ventilation. Instead of setting throughput at a fixed air replacement rate, carbon dioxide sensors are used to control the rate dynamically, based on the emissions of actual building occupants.

For the past several years, there have been many debates among indoor air quality specialists about the proper definition of indoor air quality and specifically what constitutes "acceptable" indoor air quality.

One way of quantitatively ensuring the health of indoor air is by the frequency of effective turnover of interior air by replacement with outside air. In the UK, for example, classrooms are required to have 2.5 outdoor air changes per hour. In halls, gym, dining, and physiotherapy spaces, the ventilation should be sufficient to limit carbon dioxide to 1,500 ppm. In the USA, and according to ASHRAE Standards, ventilation in classrooms is based on the amount of outdoor air per occupant plus the amount of outdoor air per unit of floor area, not air changes per hour. Since carbon dioxide indoors comes from occupants and outdoor air, the adequacy of ventilation per occupant is indicated by the concentration indoors minus the concentration outdoors. The value of 615 ppm above the outdoor concentration indicates approximately 15 cubic feet per minute of outdoor air per adult occupant doing sedentary office work where outdoor air contains 385 ppm, the current global average atmospheric CO_2 concentration. In classrooms, the requirements in the ASHRAE standard 62.1, Ventilation for Acceptable Indoor Air Quality, would typically result in about 3 air changes per hour, depending on the occupant density. Of course the occupants aren't the only source of pollutants, so outdoor air ventilation may need to be higher when unusual or strong sources of pollution exist indoors. When outdoor air is polluted, then bringing in more outdoor air can actually worsen the overall quality of the indoor air and exacerbate some occupant symptoms related to outdoor air pollution. Generally, outdoor country air is better than indoor city air. Exhaust gas leakages can occur from furnace metal exhaust pipes that lead to the chimney when there are leaks in the pipe and the pipe gas flow area diameter has been reduced.

The use of air filters can trap some of the air pollutants. The Department of Energy's Energy Efficiency and Renewable Energy section wrote "[Air] Filtration should have a Minimum Efficiency Reporting Value (MERV) of 13 as determined by ASHRAE 52.2-1999." Air filters are used to reduce the amount of dust that reaches the wet coils. Dust can serve as food to grow molds on the wet coils and ducts and can reduce the efficiency of the coils.

Moisture management and humidity control requires operating HVAC systems as designed. Moisture management and humidity control may conflict with efforts to try to optimize the operation to conserve energy. For example, Moisture management and humidity control requires systems to be set to supply Make Up Air at lower temperatures (design levels), instead of the higher temperatures sometimes used to conserve energy in cooling-dominated climate conditions. However, for most of the US and many parts of Europe and Japan, during the majority of hours of the year, outdoor air temperatures are cool enough that the air does not need further cooling to provide thermal comfort indoors. However, high humidity outdoors creates the need for careful attention to humidity levels indoors. High humidities give rise to mold growth and moisture indoors is associated with a higher prevalence of occupant respiratory problems.

The "dew point temperature" is an absolute measure of the moisture in air. Some facilities are being designed with the design dew points in the lower 50s °F, and some in the upper and lower 40s °F. Some facilities are being designed using desiccant wheels with gas fired heater to dry out the wheel enough to get the required dew points. On those systems, after the moisture is removed from the make up air, a cooling coil is used to lower the temperature to the desired level.

Commercial buildings, and sometimes residential, are often kept under slightly positive air pressure relative to the outdoors to reduce infiltration. Limiting infiltration helps with moisture management and humidity control.

Dilution of indoor pollutants with outdoor air is effective to the extent that outdoor air is free of harmful pollutants. Ozone in outdoor air occurs indoors at reduced concentrations because ozone is highly reactive with many chemicals found indoors. The products of the reactions between ozone and many common indoor pollutants include organic compounds that may be more odorous, irritating, or toxic than those from which they are formed. These products of ozone chemistry include formaldehyde, higher molecular weight aldehydes, acidic aerosols, and fine and ultrafine particles, among others. The higher the outdoor ventilation rate, the higher the indoor ozone concentration and the more likely the reactions will occur, but even at low levels, the reactions will take place. This suggests that ozone should be removed from ventilation air, especially in areas where outdoor ozone levels are frequently high. Recent research has shown that mortality and morbidity increase in the general population during periods of higher outdoor ozone and that the threshold for this effect is around 20 parts per billion (ppb).

Building Ecology

It is common to assume that buildings are simply inanimate physical entities, relatively stable over time. This implies that there is little interaction between the triad of the building, what is in it (occupants and contents), and what is around it (the larger environment). We commonly see the overwhelming majority of the mass of material in a building as relatively unchanged physical material over time. In fact, the true nature of buildings can be viewed as the result of a complex set of dynamic interactions among their physical, chemical, and biological dimensions. Buildings can be described and understood as complex systems. Research applying the approaches ecologists use to the understanding of ecosystems can help increase our understanding. "Building ecology " is proposed here as the application of those approaches to the built environment considering the dynamic system of buildings, their occupants, and the larger environment.

Buildings constantly evolve as a result of the changes in the environment around them as well as the occupants, materials, and activities within them. The various surfaces and the air inside a building are constantly interacting, and this interaction results in changes in each. For example, we may see a window as changing slightly over time as it becomes dirty, then is cleaned, accumulates dirt again, is cleaned again, and so on through its life. In fact, the "dirt" we see may be evolving as a result of the interactions among the moisture, chemicals, and biological materials found there.

Buildings are designed or intended to respond actively to some of these changes in and around them with heating, cooling, ventilating, air cleaning or illuminating systems. We clean, sanitize, and maintain surfaces to enhance their appearance, performance, or longevity. In other cases, such changes subtly or even dramatically alter buildings in ways that may be important to their own integrity or their impact on building occupants through the evolution of the physical, chemical, and biological processes that define them at any time. We may find it useful to combine the tools of the physical sciences with those of the biological sciences and, especially, some of the approaches used by scientists studying ecosystems, in order to gain an enhanced understanding of the environments in which we spend the majority of our time, our buildings.

Building ecology was first described by Hal Levin in an article in the April 1981 issue of Progressive Architecture magazine. A longer discussion of Building ecology can be found at and extensive resources can be found on the Building Ecology web site Building ecology.com.

Institutional Programs

The topic of IAQ has become popular due to the greater awareness of health problems caused by mold and triggers to asthma and allergies. In the US, awareness has also been increased by the involvement of the United States Environmental Protection Agency, who have developed an "IAQ Tools for Schools" program to help improve the indoor environmental conditions in educational institutions. The National Institute for Occupational Safety and Health conducts Health Hazard Evaluations (HHEs) in workplaces at the request of employees, authorised representative of employees, or employers, to determine whether any substance normally found in the place of employment has potentially toxic effects, including indoor air quality.

A variety of scientists work in the field of indoor air quality including chemists, physicists, mechanical engineers, biologists, bacteriologists and computer scientists. Some of these professionals are certified by organisations such as the American Industrial Hygiene Association, the American Indoor Air Quality Council and the Indoor Environmental Air Quality Council.

On the international level, the International Society of Indoor Air Quality and Climate (ISIAQ), formed in 1991, organises two major conferences, the Indoor Air and the Healthy Buildings series. ISIAQ's journal *Indoor Air* is published 6 times a year and contains peer-reviewed scientific papers with an emphasis on interdisciplinary studies including exposure measurements, modeling, and health outcomes.

Smog

Smog in New York City as viewed from the World Trade Center in 1978

Smog is a type of air pollutant. The word "smog" was coined in the early 20th century as a portmanteau of the words smoke and fog to refer to smoky fog. The word was then intended to refer to what was sometimes known as pea soup fog, a familiar and serious problem in London from the 19th century to the mid 20th century. This kind of visible air pollution is composed of nitrogen oxides, sulfur oxides, ozone, smoke or particulates among others (less visible pollutants include carbon monoxide, CFCs and radioactive sources). Man-made smog is derived from coal emissions, vehicular emissions, industrial emissions, forest and agricultural fires and photochemical reactions of these emissions.

German road sign until 2008, *Verkehrsverbot bei Smog* (No traffic allowed in smog conditions)

Modern smog, as found for example in Los Angeles, is a type of air pollution derived from vehicular emission from internal combustion engines and industrial fumes that react in the atmosphere with sunlight to form secondary pollutants that also combine with the primary emissions to form photochemical smog. In certain other cities, such as Delhi, smog severity is often aggravated by stubble burning in neighboring agricultural areas. The atmospheric pollution levels of Los Angeles, Beijing, Delhi, Mexico City, Tehran and other cities are increased by inversion that traps pollution close to the ground. It is usually highly toxic to humans and can cause severe sickness, shortened life or death.

Etymology

Coinage of the term "smog" is generally attributed to Dr. Henry Antoine Des Voeux in his 1905 paper, "Fog and Smoke" for a meeting of the Public Health Congress. The July 26, 1905 edition of the London newspaper *Daily Graphic* quoted Des Voeux, "He said it required no science to see that there was something produced in great cities which was not found in the country, and that was smoky fog, or what was known as 'smog.'" The following day the newspaper stated that "Dr. Des Voeux did a public service in coining a new word for the London fog." However, this is predated by a *Los Angeles Times* article of January 19, 1893, in which the word is attributed to "a witty English writer."

Causes

Coal

Coal fires, used to heat individual buildings or in a power-producing plant, can emit significant clouds of smoke that contributes to smog. Air pollution from this source has been reported in England since the Middle Ages. London, in particular, was notorious up through the mid-20th century for its coal-caused smogs, which were nicknamed 'pea-soupers.' Air pollution of this type is still a problem in areas that generate significant smoke from burning coal, as witnessed by the 2013 autumnal smog in Harbin, China, which closed roads, schools, and the airport.

Transportation Emissions

Traffic emissions – such as from trucks, buses, and automobiles – also contribute. Airborne by-products from vehicle exhaust systems cause air pollution and are a major ingredient in the creation of smog in some large cities.

The major culprits from transportation sources are carbon monoxide (CO), nitrogen oxides (NO and NO_x), volatile organic compounds, sulfur dioxide, and hydrocarbons. (Hydrocarbons are the main components of petroleum fuels such as gasoline and diesel fuel.) These molecules react with sunlight, heat, ammonia, moisture, and other compounds to form the noxious vapors, ground level ozone, and particles that comprise smog.

Photochemical Smog

Photochemical smog is the chemical reaction of sunlight, nitrogen oxides and volatile organic compounds in the atmosphere, which leaves airborne particles and ground-level ozone. This noxious mixture of air pollutants may include the following:

- Aldehydes
- Nitrogen oxides, particularly nitric oxide and nitrogen dioxide
- Peroxyacyl nitrates
- Tropospheric ozone
- Volatile organic compounds

A primary pollutant is an air pollutant emitted directly from a source. A secondary pollutant is not directly emitted as such, but forms when other pollutants (primary pollutants) react in the atmosphere. Examples of a secondary pollutant include ozone, which is formed when hydrocarbons (HC) and nitrogen oxides (NO_x) combine in the presence of sunlight; nitrogen dioxide (NO_2), which is formed as nitric oxide (NO) combines with oxygen in the air; and acid rain, which is formed when sulfur dioxide or nitrogen oxides react with water. All of these harsh chemicals are usually highly reactive and oxidizing. Photochemical smog is therefore considered to be a problem of modern industrialization. It is present in all modern cities, but it is more common in cities with sunny, warm, dry climates and a large number of motor vehicles. Because it travels with the wind, it can affect sparsely populated areas as well.

The link between automotive exhaust and photochemical smog was discovered in the 1950s by Arie Haagen-Smit.

Characteristic coloration for smog in California in the beige cloud bank behind the Golden Gate Bridge. The brown coloration is due to the NO_x in the photochemical smog.

Natural Causes

An erupting volcano can also emit high levels of sulphur dioxide along with a large quantity of particulate matter; two key components to the creation of smog. However, the smog created as a result of a volcanic eruption is often known as vog to distinguish it as a natural occurrence.

The radiocarbon content of some plant life has been linked to the distribution of smog in some areas. For example, the creosote bush in the Los Angeles area has been shown to have an effect on smog distribution that is more than fossil fuel combustion alone.

Health Effects

Highland Park Optimist Club wearing smog-gas masks at banquet, Los Angeles, circa 1954

Smog is a serious problem in many cities and continues to harm human health. Ground-level ozone, sulfur dioxide, nitrogen dioxide and carbon monoxide are especially harmful for senior citizens, children, and people with heart and lung conditions such as emphysema, bronchitis, and asthma. It can inflame breathing passages, decrease the lungs' working capacity, cause shortness of breath, pain when inhaling deeply, wheezing, and coughing. It can cause eye and nose irritation and it dries out the protective membranes of the nose and throat and interferes with the body's ability to fight infection, increasing susceptibility to illness. Hospital admissions and respiratory deaths often increase during periods when ozone levels are high.

Levels of Unhealthy Exposure

The U.S. EPA has developed an Air Quality Index to help explain air pollution levels to the general public. 8 hour average ozone concentrations of 85 to 104 ppbv are described as "Unhealthy for Sensitive Groups", 105 ppbv to 124 ppbv as "unhealthy" and 125 ppb to 404 ppb as "very unhealthy". The "very unhealthy" range for some other pollutants are: 355 $\mu g\ m^{-3}$ - 424 $\mu g\ m^{-3}$ for PM10; 15.5 ppm - 30.4ppm for CO and 0.65 ppm - 1.24 ppm for NO_2.

Premature Deaths Due to Cancer and Respiratory Disease

The Ontario Medical Association announced that smog is responsible for an estimated 9,500 premature deaths in the province each year.

A 20-year American Cancer Society study found that cumulative exposure also increases the likelihood of premature death from a respiratory disease, implying the 8-hour standard may be insufficient.

Smog and the Risk of Certain Birth Defects

A study examining 806 women who had babies with birth defects between 1997 and 2006, and 849 women who had healthy babies, found that smog in the San Joaquin Valley area of California was linked to two types of neural tube defects: spina bifida (a condition involving, among other manifestations, certain malformations of the spinal column), and anencephaly (the underdevelopment or absence of part or all of the brain, which if not fatal usually results in profound impairment).

Smog and Low Birth Weight

According to a study published in The Lancet, even a very small (5 µg) change in PM2.5 exposure was associated with an increase (18%) in risk of a low birth weight at delivery, and this relationship held even below the current accepted safe levels.

Areas Affected

Beijing air on a day after rain (left) and a smoggy day (right)

Smog can form in almost any climate where industries or cities release large amounts of air pollution, such as smoke or gases. However, it is worse during periods of warmer, sunnier weather when the upper air is warm enough to inhibit vertical circulation. It is especially prevalent in geologic basins encircled by hills or mountains. It often stays for an extended period of time over densely populated cities or urban areas, and can build up to dangerous levels.

Delhi, India

Delhi is the most polluted city in the world and according to one estimate, air pollution causes the death of about 10,500 people in Delhi every year. During 2013-14, peak levels of fine particulate matter (PM) in Delhi increased by about 44%, primarily due to high vehicular and industrial emissions, construction work and crop burning in adjoining states. Delhi has the highest level of the airborne particulate matter, PM2.5 considered most harmful to health, with 153 micrograms. Rising air pollution level has significantly increased lung-related ailments (especially asthma and lung cancer) among Delhi's children and women. The dense smog in Delhi during winter season results in major air and rail traffic disruptions every year. According to Indian meteorologists, the average maximum temperature in Delhi during winters has declined notably since 1998 due to rising air pollution.

During the autumn and winter months, some 500 million tons of crop residue are burnt, and winds blow from India's north and northwest towards east. This aerial view shows India's annual crop burning, resulting in smoke and air pollution over Delhi and adjoining areas.

Environmentalists have criticised the Delhi government for not doing enough to curb air pollution and to inform people about air quality issues. Most of Delhi's residents are unaware of alarming levels of air pollution in the city and the health risks associated with it. Since the mid-1990s, Delhi has undertaken some measures to curb air pollution – Delhi has the third highest quantity of trees among Indian cities and the Delhi Transport Corporation operates the world's largest fleet of environmentally friendly compressed natural gas (CNG) buses. In 1996, the Centre for Science and Environment (CSE) started a public interest litigation in the Supreme Court of India that ordered the conversion of Delhi's fleet of buses and taxis to run on CNG and banned the use of leaded petrol in 1998. In 2003, Delhi won the United States Department of Energy's first 'Clean Cities International Partner of the Year' award for its "bold efforts to curb air pollution and support alternative fuel initiatives". The Delhi Metro has also been credited for significantly reducing air pollutants in the city.

Dense smog blankets Connaught Place, New Delhi.

However, according to several authors, most of these gains have been lost, especially due to stubble burning, rise in market share of diesel cars and a considerable decline in bus ridership. According to CSE and System of Air Quality Weather Forecasting and Research (SAFAR), burning of agricultural waste in nearby Punjab, Haryana and Uttar Pradesh regions results in severe intensification of smog over Delhi. The state government of adjoining Uttar Pradesh is considering imposing a ban on crop burning to reduce pollution in Delhi NCR and an environmental panel has appealed to India's Supreme Court to impose a 30% cess on diesel cars.

United Kingdom

London

Victorian London was notorious for its thick smogs, or "pea-soupers", a fact that is often recreated (as here) to add an air of mystery to a period costume drama

In 1306, concerns over air pollution were sufficient for Edward I to (briefly) ban coal fires in London. In 1661, John Evelyn's *Fumifugium* suggested burning fragrant wood instead of mineral coal, which he believed would reduce coughing. The "Ballad of Gresham College" the same year describes how the smoke "does our lungs and spirits choke, Our hanging spoil, and rust our iron."

Severe episodes of smog continued in the 19th and 20th centuries, mainly in the winter, and were nicknamed "pea-soupers," from the phrase "as thick as pea soup." The Great Smog of 1952 darkened the streets of London and killed approximately 4,000 people in the short time of 4 days (a further 8,000 died from its effects in the following weeks and months). Initially a flu epidemic was blamed for the loss of life.

In 1956 the Clean Air Act started legally enforcing smokeless zones in the capital. There were areas where no soft coal was allowed to be burned in homes or in businesses, only coke, which produces no smoke. Because of the smokeless zones, reduced levels of sooty particulates eliminated the intense and persistent London smog.

It was after this that the great clean-up of London began. One by one, historical buildings which, during the previous two centuries had gradually completely blackened externally, had their stone facades cleaned and restored to their original appearance. Victorian buildings whose appearance changed dramatically after cleaning included the British Museum of Natural History. A more recent example was the Palace of Westminster, which was cleaned in the 1980s. A notable exception to the restoration trend was 10 Downing Street, whose bricks upon cleaning in the late 1950s proved to be naturally *yellow*; the smog-derived black colour of the façade was considered so iconic that the bricks were painted black to preserve the image. Smog caused by traffic pollution, however, does still occur in modern London.

Other Areas

Other areas of the United Kingdom were affected by smog, especially heavily industrialised areas.

The cities of Glasgow and Edinburgh, in Scotland, suffered smoke-laden fogs in 1909. Des Voeux, commonly credited with creating the "smog" moniker, presented a paper in 1911 to the Manchester Conference of the Smoke Abatement League of Great Britain about the fogs and resulting deaths.

One Birmingham resident described near black-out conditions in the 1900s before the Clean Air Act, with visibility so poor that cyclists had to dismount and walk in order to stay on the road.

Mexico City, Mexico

Situated in a valley, and relying heavily on automobiles, Mexico City often suffers from poor air quality.

Photochemical smog over Mexico City. December 2010.

Due to its location in a highland "bowl", cold air sinks down onto the urban area of Mexico City, trapping industrial and vehicle pollution underneath, and turning it into the most infamously smog-plagued city of Latin America. Within one generation, the city has changed from being known for some of the cleanest air of the world into one with some of the worst pollution, with pollutants like nitrogen dioxide being double or even triple international standards.

Santiago, Chile

Similar to Mexico City, the air pollution of Santiago valley, located between the Andes and the Chilean Coast Range, turn it into the most infamously smog-plagued city of South America. Other aggravates of the situation reside in its high latitude (31 degrees South) and dry weather during most of the year.

Tehran, Iran

In December 2005, schools and public offices had to close in Tehran, Iran and 1600 people were taken to hospital, in a severe smog blamed largely on unfiltered car exhaust.

United States

A NASA astronaut photograph of a smog layer over central New York.

View of smog south from Los Angeles City Hall, September 2011.

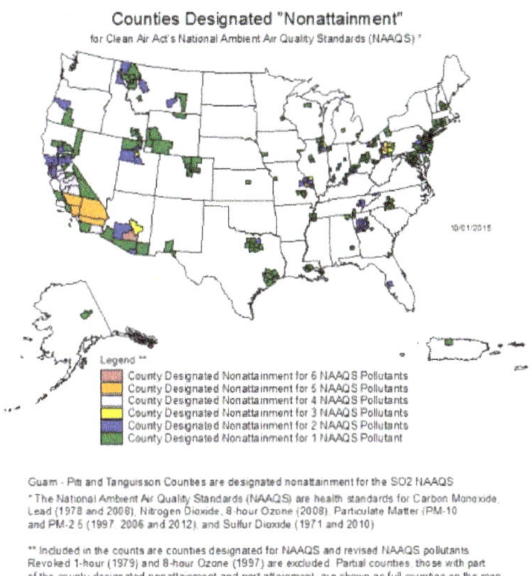

Counties in the United States where one or more National Ambient Air Quality Standards are not met, as of October 2015.

Smog was brought to the attention of the general US public in 1933 with the publication of the book "Stop That Smoke", by Henry Obermeyer, a New York public utility official, in which he pointed out the effect on human life and even the destruction of 3,000 acres (12 km^2) of a farmer's spinach crop. Since then, the United States Environmental Protection Agency has designated over 300 U.S. counties to be non-attainment areas for one or more pollutants tracked as part of the National Ambient Air Quality Standards. These areas are largely clustered around large metropolitan areas, with the largest contiguous non-attainment zones in California and the Northeast. Various U.S. and Canadian government agencies collaborate to produce real-time air quality maps and forecasts.

Los Angeles and the San Joaquin Valley

Because of their locations in low basins surrounded by mountains, Los Angeles and the San Joaquin Valley are notorious for their smog. The millions of vehicles in these regions combined with the additional effects of the San Francisco Bay and Los Angeles/Long Beach port complexes frequently contribute to further air pollution.

Los Angeles in particular is strongly predisposed to accumulation of smog, because of peculiarities of its geography and weather patterns. Los Angeles is situated in a flat basin with ocean on one side and mountain ranges on three sides. A nearby cold ocean current depresses surface air temperatures in the area, resulting in an inversion layer: a phenomenon where air temperature increases, instead of decreasing, with altitude, suppressing thermals and restricting vertical convection. All taken together, this results in a relatively thin, enclosed layer of air above the city that cannot easily escape out of the basin and tends to accumulate pollution.

Though Los Angeles was one of the best known cities suffering from transportation smog for much of the 20th century, so much so that it was sometimes said that *Los Angeles* was a synonym for *smog*, strict regulations by government agencies overseeing this problem, including tight restrictions on allowed emissions levels for all new cars sold in California and mandatory regular emission tests of older vehicles, resulted in significant improvements in air quality. For example, air concentrations of volatile organic compounds declined by a factor of 50 between 1962 and 2012. Concentrations of air pollutants such as nitrous oxides and ozone declined by 70% to 80% over the same period of time.

Major Incidents in the Us

- 1943, July 26, Los Angeles, California: A smog so sudden and severe that "Los Angeles residents believe the Japanese are attacking them with chemical warfare."
- 1948, October 30–31, Donora, Pennsylvania: 20 died, 600 hospitalized, thousands more stricken. Lawsuits were not settled until 1951.
- 1966, November 24, New York City, New York: Smog kills at least 169 people.

Ulaanbaatar, Mongolia

In the late 1990s, massive immigration to Ulaanbaatar from the countryside began. An estimated 150,000 households, mainly living in traditional Mongolian gers on the outskirts of Ulaanbaatar, burn wood and coal (some poor families burn even car tires and trash) to heat themselves during the harsh winter, which lasts from October to April, since these outskirts are not connected to the city's central heating system. A temporary solution to decrease smog was proposed in the form of stoves with improved efficiency, although with no visible results. Coal-fired ger stoves release high levels of ash and other particulate matter (PM). When inhaled, these particles can settle in the lungs and respiratory tract and cause health problems. At two to 10 times above Mongolian and international air quality standards, Ulaanbaatar's PM rates are among the worst in the world, according to a December 2009 World Bank report. The Asian Development Bank (ADB) estimates that health costs related to this air pollution account for as much as 4 percent of Mongolia's GDP.

Southeast Asia

Singapore's Downtown Core on 7 October 2006, when it was affected by forest fires in Sumatra, Indonesia

Smog is a regular problem in Southeast Asia caused by land and forest fires in Indonesia, especially Sumatra and Kalimantan, although the term haze is preferred in describing the problem. Farmers and plantation owners are usually responsible for the fires, which they use to clear tracts of land for further plantings. Those fires mainly affect Brunei, Indonesia, Philippines, Malaysia, Singapore and Thailand, and occasionally Guam and Saipan. The economic losses of the fires in 1997 have been estimated at more than US$9 billion. This includes damages in agriculture production, destruction of forest lands, health, transportation, tourism, and other economic endeavours. Not included are social, environmental, and psychological problems and long-term health effects. The second-latest bout of haze to occur in Malaysia, Singapore and the Malacca Straits is in October 2006, and was caused by smoke from fires in Indonesia being blown across the Straits of Malacca by south-westerly winds. A similar haze has occurred in June 2013, with the PSI setting a new record in Singapore on June 21 at 12pm with a reading of 401, which is in the "Hazardous" range.

The Association of Southeast Asian Nations (ASEAN) reacted. In 2002, the Agreement on Transboundary Haze Pollution was signed between all ASEAN nations. ASEAN formed a Regional Haze Action Plan (RHAP) and established a co-ordination and support unit (CSU). RHAP, with the help of Canada, established a monitoring and warning system for forest/vegetation fires and implemented a Fire Danger Rating System (FDRS). The Malaysian Meteorological Department (MMD) has issued a daily rating of fire danger since September 2003. Indonesia has been ineffective at enforcing legal policies on errant farmers.

Pollution Index

Smog in São Paulo, Brazil

The severity of smog is often measured using automated optical instruments such as Nephelometers, as haze is associated with visibility and traffic control in ports. Haze however can also be an indication of poor air quality though this is often better reflected using accurate purpose built air indexes such as the American Air Quality Index, the Malaysian API (Air Pollution Index) and the Singaporean Pollutant Standards Index.

In hazy conditions, it is likely that the index will report the suspended particulate level. The disclosure of the responsible pollutant is mandated in some jurisdictions.

The Malaysian API does not have a capped value; hence its most hazardous readings can go above 500. Above 500, a state of emergency is declared in the affected area. Usually, this means that non-essential government services are suspended, and all ports in the affected area are closed. There may also be prohibitions on private sector commercial and industrial activities in the affected area excluding the food sector. So far, state of emergency rulings due to hazardous API levels were applied to the Malaysian towns of Port Klang, Kuala Selangor and the state of Sarawak during the 2005 Malaysian haze and the 1997 Southeast Asian haze.

Cultural References

Claude Monet made several trips to London between 1899 and 1901, during which he painted views of the Thames and Houses of Parliament which show the sun struggling to shine through London's smog-laden atmosphere.

- The London "pea-soupers" earned the capital the nickname of "The Smoke". Similarly, Edinburgh was known as "Auld Reekie". The smogs feature in many London novels as a motif indicating hidden danger or a mystery, perhaps most overtly in Margery Allingham's *The Tiger in the Smoke* (1952), but also in Dickens's *Bleak House* (1852) and T.S. Eliot's "The Love Song of J. Alfred Prufrock."
- The 1970 made-for-TV movie *A Clear and Present Danger* was one of the first American television network entertainment programs to warn about the problem of smog and air pollution, as it dramatized a man's efforts toward clean air after emphysema killed his friend.
- The history of smog in LA is detailed in *Smogtown* by Chip Jacobs and William J. Kelly.

Neuroplastic Effects of Pollution

Research indicates that living in areas of high pollution has serious long term health effects. Living in these areas during childhood and adolescence can lead to diminished mental capacity and an increased risk of brain damage. People of all ages who live in high pollution areas for extended periods place themselves at increased risk of various neurological disorders. Both air pollution and heavy metal pollution have been implicated as having negative effects on central nervous system (CNS) functionality. The ability of pollutants to affect the neurophysiology of individuals after the structure of the CNS has become mostly stabilized is an example of negative neuroplasticity.

Air Pollution

Air pollution is known to affect small and large blood vessels throughout the body. High levels of air pollution are associated with increased risk of strokes and heart attacks. By permanently affecting vascular structures in the brain, air pollution can have serious effects on neural functioning and neural matter. In dogs air pollution shows to cause damage to the CNS by altering the blood–brain barrier, causing neurons in the cerebral cortex to degenerate, destroying glial cells found in white matter, and by causing neurofibrillary tangles. These changes can permanently alter brain structure and chemistry, resulting in various impairments and disorders. Sometimes, the effects of neural remodeling do not manifest themselves for a prolonged period of time.

Effects in Adolescents and Canines

A study from 2008 compared children and dogs raised in Mexico City (a location known for high pollution levels) with children and dogs raised in Polotitlán, Mexico (a city whose pollution levels meet the current US National Ambient Air Quality Standards). According to this study, children raised in areas of higher pollution scored lower in intelligence (i.e. on IQ tests), and showed signs of lesions in MRI scanning of the brain. In contrast, children from the low pollution area scored as expected on IQ tests, and did not show any significant sign of the risk of brain lesions. This correlation was found to be statistically significant, and shows that pollution levels may be related to, and contribute to, brain lesion formation and IQ scores, which, in turn, manifests as impaired intellectual capacity and/or performance. Living in high pollution areas thus places adolescents at risk of premature brain degeneration and improper neural development—these findings could have significant implications for future generations.

Effects in Adults

There are indications that the effects of physical activity and air pollution on neuroplasticity counteract. Physical activity is known for its health-enhancing benefits, particularly on the cardiovascular system, and has also demonstrated benefits for brain plasticity processes, cognition and mental health. The neurotrophine, brain-derived neurotrophic factor (BDNF) is thought to play a key role in exercise-induced cognitive improvements. Brief bouts of physical activity have been shown to increase serum levels of BDNF, but this increase may be offset by increased exposure to traffic-related air pollution. Over longer periods of physical exercise, cognitive improvements that were demonstrated in rural joggers were found to be absent in urban joggers taking the same 12-week start-2-run training programme.

Epilepsy

Researchers in Chile found statistically-significant correlations between multiple air pollutants and the risk of epilepsy using a 95% confidence interval. The air pollutants that the researchers attempted to correlate with increased incidence of epilepsy included carbon monoxide, ozone, sulfur dioxide, nitrogen dioxide, large particulate matter, and fine particulate matter. The researchers tested these pollutants across seven cities and, in all but one case, a correlation was found between pollutant levels and the occurrence of epilepsy. Interestingly, all of the correlations found were shown to be statistically significant. The researchers hypothesized that air pollutants increase epilepsy risk by increasing inflammatory mediators, and by providing a source of oxidative stress.

They believe that these changes eventually alter the functioning of the blood–brain barrier, causing brain inflammation. Brain inflammation is known to be a risk factor for epilepsy; thus, the sequence of events provides a plausible mechanism by which pollution may increase epilepsy risk in individuals who are genetically vulnerable to the disease.

Dioxin Poisoning

Organohalogen compounds, such as dioxins, are commonly found in pesticides or created as by-products of pesticide manufacture or degradation. These compounds can have a significant impact on the neurobiology of exposed organisms. Some observed effects of exposure to dioxins are altered astroglial intracellular calcium ion (Ca^{2+}), decreased glutathione levels, modified neurotransmitter function in the CNS, and loss of pH maintenance. A study of 350 chemical plant employees exposed to a dioxin precursor for herbicide synthesis between 1965 and 1968 showed that 80 of the employees displayed signs of dioxin poisoning. Of these 350 employees, 15 were contacted again in 2004 to submit to neurological tests to assess whether the dioxin poisoning had any long-term effects on neurological capabilities. The amount of time that had passed made it difficult to assemble a larger cohort, but the results of the tests indicated that eight of the 15 subjects exhibited some central nervous system impairment, nine showed signs of polyneuropathy, and electroencephalography (EEG) showed various degrees of structural abnormalities. This study suggested that the effects of dioxins were not limited to initial toxicity. Dioxins, through neuroplastic effects, can cause long-term damage that may not manifest itself for years or even decades.

Metal Exposure

Heavy metal exposure can result in an increased risk of various neurological diseases. Research indicates that the two most neurotoxic heavy metals are mercury and lead. The impact that these two metals will have is highly dependent upon the individual due to genetic variations. Mercury and lead are particularly neurotoxic for many reasons: they easily cross cell membranes, have oxidative effects on cells, react with sulfur in the body (leading to disturbances in the many functions that rely upon sulfhydryl groups), and reduce glutathione levels inside cells. Methylmercury, in particular, has an extremely high affinity for sulfhydryl groups. Organomercury is a particularly damaging form of mercury because of its high absorbability Lead also mimics calcium, a very important mineral in the CNS, and this mimicry leads to many adverse effects. Mercury's neuroplastic mechanisms work by affecting protein production. Elevated mercury levels increase glutathione levels by affecting gene expression, and this in turn affects two proteins (MT1 and MT2) that are contained in astrocytes and neurons. Lead's ability to imitate calcium allows it to cross the blood–brain barrier. Lead also upregulates glutathione.

Autism

Heavy metal exposure, when combined with certain genetic predispositions, can place individuals at increased risk for developing autism. Many examples of CNS pathophysiology, such as oxidative stress, neuroinflammation, and mitochondrial dysfunction, could be by-products of environmental stressors such as pollution. There have been reports of autism outbreaks occurring in specific locations. Since these cases of autism are related to geographic location, the implication is that something in the environment is complementing an at-risk genotype to cause autism in these vulnerable individuals. Mercury and lead both contribute to inflammation, leading scientists to speculate that these heavy metals could play a role in autism. These findings are controversial, however, with many researchers believing that increasing rates of autism are a consequence of more accurate screening and diagnostic methods, and are not due to any sort of environmental factor.

Accelerated Neural Aging

Neuroinflammation is associated with increased rates of neurodegeneration. Inflammation tends to increase naturally with age. By facilitating inflammation, pollutants such as air particulates and heavy metals cause the CNS to age more quickly. Many late-onset diseases are caused by neurodegeneration. Multiple sclerosis, Parkinson's disease, amyotrophic lateral sclerosis (ALS), and Alzheimer's disease are all believed to be exacerbated by inflammatory processes, resulting in individuals displaying signs of these diseases at an earlier age than is typically expected.

Multiple sclerosis occurs when chronic inflammation leads to the compromise of oligodendrocytes, which in turn leads to the destruction of the myelin sheath. Then axons begin exhibiting signs of damage, which in turn leads to neuron death. Multiple sclerosis has been correlated to living in areas with high particulate matter levels in the air.

In Parkinson's disease, inflammation leading to depletion of antioxidant stores will ultimately lead to dopaminergic neuron degeneration, causing a shortage of dopamine and contributing to the formation of Parkinson's disease. Chronic glial activation as a result of inflammation causes motor neuron death and compromises astrocytes, these factors leading to the symptoms of amyotrophic lateral sclerosis (ALS, aka Lou Gehrig's disease).

In the case of Alzheimer's disease, inflammatory processes lead to neuron death by inhibiting growth at axons and activating astrocytes that produce proteoglycans. This product can only be deposited in the hippocampus and cortex, indicating that this may be the reason these two areas show the highest levels of degeneration in Alzheimer's disease. Airborne metal particulates have been shown to directly access and affect the brain through olfactory pathways, which allows a large amount of particulate matter to reach the blood–brain barrier.

These facts, coupled with air pollution's link to neurofibrillary tangles and the observed subcortical vascular changes observed in dogs, imply that the negative neuroplastic effects of pollution could result in increased risk for Alzheimer's disease, and could also implicate pollution as a cause of early-onset Alzheimer's disease through multiple mechanisms. The general effect of pollution is increased levels of inflammation. As a result, pollution can significantly contribute to various neurological disorders that are caused by inflammatory processes.

References

- May, Jeffrey C.; Ouellette, Connie L. May ; with a contribution by John J., Reed, Charles E. (2004). The mold survival guide for your home and for your health. Baltimore: Johns Hopkins University Press. ISBN 978-0-8018-7938-8.
- May, Jeffrey C. (2001). My house is killing me! : the home guide for families with allergies and asthma. Baltimore: The Johns Hopkins University Press. ISBN 978-0-8018-6730-9.
- May, Jeffrey C. (2006). My office is killing me! : the sick building survival guide. Baltimore: The Johns Hopkins University Press. ISBN 978-0-8018-8342-2.

- Salthammer, T., ed. (1999). Organic Indoor Air Pollutants — Occurrence, Measurement, Evaluation. Wiley-VCH. ISBN 3-527-29622-0.
- Spengler, J.D., Samet, J.M. (1991). Indoor air pollution: A health perspective. Baltimore: Johns Hopkins University Press. ISBN 0-8018-4125-9.
- Tichenor, B. (1996). Characterizing Sources of Indoor Air Pollution and Related Sink Effects. ASTM STP 1287. West Conshohocken, PA: ASTM. ISBN 0-8031-2030-3.
- Miller, Jr., George Tyler (2002). Living in the Environment: Principles, Connections, and Solutions (12th Edition). Belmont: The Thomson Corporation. p. 423. ISBN 0-534-37697-5.
- Buntin, John (2009). L.A. Noir: The Struggle for the Soul of America's Most Seductive City. New York: Harmony Books. p. 108. ISBN 9780307352071. OCLC 431334523. Retrieved 12 October 2014.
- Jacobs, Chip; Kelly, William J. (4 October 2009). Smogtown, The Lung-Burning History of Pollution in Los Angeles. Overlook Press. ISBN 978-1-58567-860-0.
- Gardner, Sarah (14 July 2014). "LA Smog: the battle against air pollution". Marketplace.org. American Public Media. Retrieved 6 November 2015.
- Tracton, Steve (December 20, 2012). "The Killer London Smog Event of December 1952: A Reminder of Deadly Smog Events in the US". The Washington Post. Retrieved February 25, 2015.
- "Delhi's Air Has Become a Lethal Hazard and Nobody Seems to Know What to Do About It". Time magazine. Retrieved 10 February 2014.
- HARRIS, GARDINER (25 January 2014). "Beijing's Bad Air Would Be Step Up for Smoggy Delhi". New York Times. Retrieved 27 January 2014.
- BEARAK, MAX (7 February 2014). "Desperate for Clean Air, Delhi Residents Experiment with Solutions". New York Times. Retrieved 8 February 2014.

5

Ozone Depletion: A Global Concern

Air pollution can be best understood in confluence with the major topics listed in the following chapter. The ozone layer blankets the earth and filters harmful ultra violent radiation and is responsible for maintaining global temperature. The rays emitted back to earth causes skin and eye related problems and also has the capability to affect crops. This topic is of great importance to have a better understanding of air pollution.

Ozone depletion describes two distinct but related phenomena observed since the late 1970s: a steady decline of about four percent in the total amount of ozone in Earth's stratosphere (the ozone layer), and a much larger springtime decrease in stratospheric ozone around Earth's polar regions. The latter phenomenon is referred to as the ozone hole. In addition to these well-known stratospheric phenomena, there are also springtime polar tropospheric ozone depletion events.

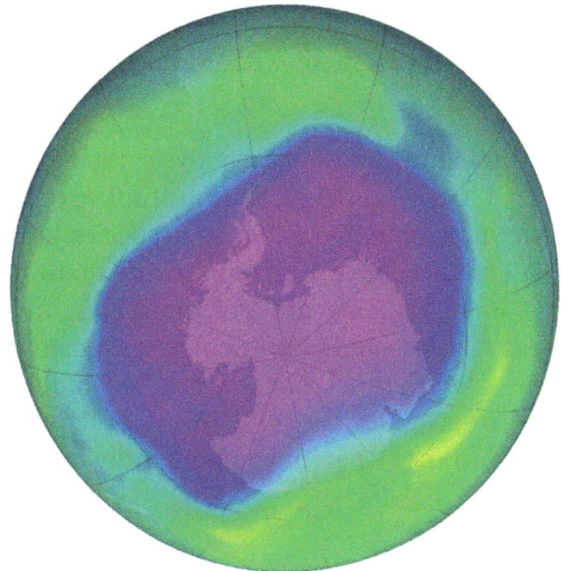

Image of the largest Antarctic ozone hole ever recorded (September 2006), over the Southern pole

The details of polar ozone hole formation differ from that of mid-latitude thinning but the most important process in both is catalytic destruction of ozone by atomic halogens. The main source of these halogen atoms in the stratosphere is photodissociation of man-made halocarbon refrigerants, solvents, propellants, and foam-blowing agents (chlorofluorocarbon (CFCs), HCFCs, freons, halons). These compounds are transported into the stratosphere by winds after being emitted at the surface. Both types of ozone depletion were observed to increase as emissions of halocarbons increased.

CFCs and other contributory substances are referred to as ozone-depleting substances (ODS). Since the ozone layer prevents most harmful UVB wavelengths (280–315 nm) of ultraviolet light (UV light) from passing through the Earth's atmosphere, observed and projected decreases in ozone generated worldwide concern, leading to adoption of the Montreal Protocol that bans the production of CFCs, halons, and other ozone-depleting chemicals such as carbon tetrachloride and trichloroethane. It is suspected that a variety of biological consequences such as increases in sunburn, skin cancer, cataracts, damage to plants, and reduction of plankton populations in the ocean's photic zone may result from the increased UV exposure due to ozone depletion.

Ozone Cycle Overview

Three forms (or allotropes) of oxygen are involved in the ozone-oxygen cycle: oxygen atoms (O or atomic oxygen), oxygen gas (O_2 or diatomic oxygen), and ozone gas (O_3 or triatomic oxygen). Ozone is formed in the stratosphere when oxygen molecules photodissociate after intaking an ultraviolet photon whose wavelength is shorter than 240 nm. This converts a single O_2 into two atomic oxygen radicals. The atomic oxygen radicals then combine with separate O_2 molecules to create two O_3 molecules. These ozone molecules absorb ultraviolet (UV) light between 310 and 200 nm, following which ozone splits into a molecule of O_2 and an oxygen atom. The oxygen atom then joins up with an oxygen molecule to regenerate ozone. This is a continuing process that terminates when an oxygen atom "recombines" with an ozone molecule to make two O_2 molecules.

$$2\,O_3 \rightarrow 3\,O_2$$

The overall amount of ozone in the stratosphere is determined by a balance between photochemical production and recombination.

Ozone can be destroyed by a number of free radical catalysts, the most important of which are the hydroxyl radical (OH·), nitric oxide radical (NO·), chlorine radical (Cl·) and bromine radical (Br·). The dot is a common notation to indicate that all of these species have an unpaired electron and are thus extremely reactive. All of these have both natural and man-made sources; at the present time, most of the OH· and NO· in the stratosphere is of natural origin, but human activity has dramatically increased the levels of chlorine and bromine. These elements are found in certain stable organic compounds, especially chlorofluorocarbons, which may find their way to the stratosphere without being destroyed in the troposphere due to their low reactivity. Once in the stratosphere, the Cl and Br atoms are liberated from the parent compounds by the action of ultraviolet light, e.g.

$CFCl_3$ + electromagnetic radiation → Cl· + ·$CFCl_2$

The Cl and Br atoms can then destroy ozone molecules through a variety of catalytic cycles. In the simplest example of such a cycle, a chlorine atom reacts with an ozone molecule, taking an oxygen atom with it (forming ClO) and leaving a normal oxygen molecule. The chlorine monoxide (i.e., the ClO) can react with a second molecule of ozone (i.e., O_3) to yield another chlorine atom and two molecules of oxygen. The chemical shorthand for these gas-phase reactions is:

- Cl· + O_3 → ClO + O_2: The chlorine atom changes an ozone molecule to ordinary oxygen
- ClO + O_3 → Cl· + 2 O_2: The ClO from the previous reaction destroys a second ozone molecule and recreates the original chlorine atom, which can repeat the first reaction and continue to destroy ozone.

The overall effect is a decrease in the amount of ozone, though the rate of these processes can be decreased by the effects of null cycles. More complicated mechanisms have been discovered that lead to ozone destruction in the lower stratosphere as well.

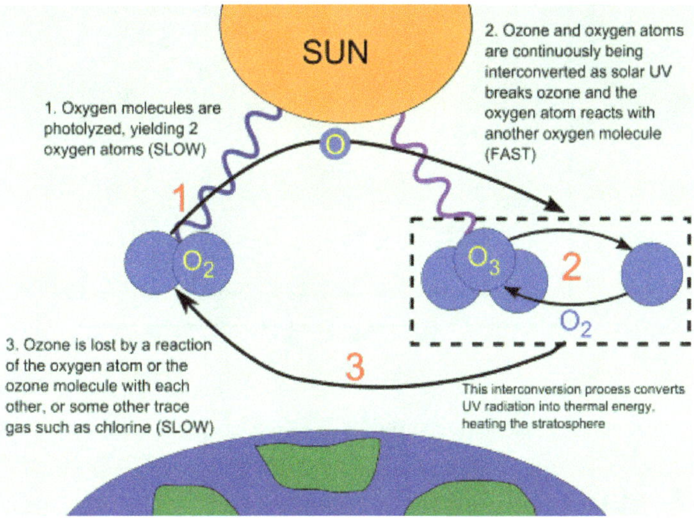

The ozone cycle

A single chlorine atom would keep on destroying ozone (thus a catalyst) for up to two years (the time scale for transport back down to the troposphere) were it not for reactions that remove them from this cycle by forming reservoir species such as hydrogen chloride (HCl) and chlorine nitrate (ClONO2). On a per atom basis, bromine is even more efficient than chlorine at destroying ozone, but there is much less bromine in the atmosphere at present. As a result, both chlorine and bromine contribute significantly to overall ozone depletion. Laboratory studies have shown that fluorine and iodine atoms participate in analogous catalytic cycles. However, in the Earth's stratosphere, fluorine atoms react rapidly with water and methane to form strongly bound HF, while organic molecules containing iodine react so rapidly in the lower atmosphere that they do not reach the stratosphere in significant quantities.

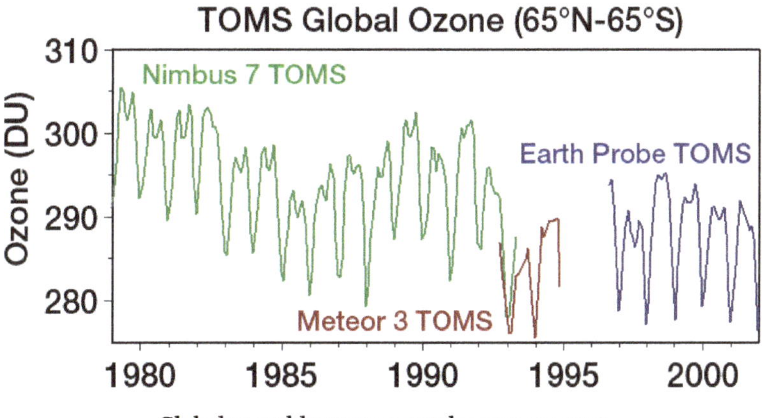

Global monthly average total ozone amount

On average, a single chlorine atom is able to react with 100,000 ozone molecules before it is removed from the catalytic cycle. This fact plus the amount of chlorine released into the atmosphere yearly by chlorofluorocarbons (CFCs) and hydrochlorofluorocarbons (HCFCs) demonstrates how dangerous CFCs and HCFCs are to the environment.

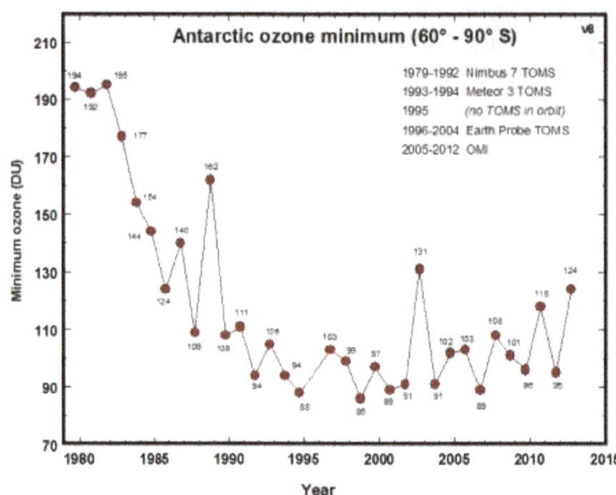

Lowest value of ozone measured by TOMS each year in the ozone hole

Observations on Ozone Layer Depletion

The most-pronounced decrease in ozone has been in the lower stratosphere. However, the ozone hole is most usually measured not in terms of ozone concentrations at these levels (which are typically a few parts per million) but by reduction in the total *column ozone* above a point on the Earth's surface, which is normally expressed in Dobson units, abbreviated as "DU". Marked decreases in column ozone in the Antarctic spring and early summer compared to the early 1970s and before have been observed using instruments such as the Total Ozone Mapping Spectrometer (TOMS).

Reductions of up to 70 percent in the ozone column observed in the austral (southern hemispheric) spring over Antarctica and first reported in 1985 (Farman et al.) are continuing. Since the 1990s, Antarctic total column ozone in September and October continued to be 40–50 percent lower than pre-ozone-hole values. A gradual trend toward "healing" was reported in 2016.

In the Arctic, the amount lost is more variable year-to-year than in the Antarctic. The greatest Arctic declines, up to 30 percent, are in the winter and spring, when the stratosphere is coldest.

Reactions that take place on polar stratospheric clouds (PSCs) play an important role in enhancing ozone depletion. PSCs form more readily in the extreme cold of the Arctic and Antarctic stratosphere. This is why ozone holes first formed, and are deeper, over Antarctica. Early models failed to take PSCs into account and predicted a gradual global depletion, which is why the sudden Antarctic ozone hole was such a surprise to many scientists.

In middle latitudes, it is more accurate to speak of ozone depletion rather than holes. Total column ozone declined to about six percent below pre-1980 values between 1980 and 1996 for the mid-latitudes of 35–60°N and 35–60°S. In the northern mid-latitudes, it then increased from the minimum value by about two percent from 1996 to 2009 as regulations took effect and the amount of chlorine in the stratosphere decreased. In the Southern Hemisphere's mid-latitudes, total ozone remained constant over that time period. In the tropics, there are no significant trends, largely because halogen-containing compounds have not yet had time to break down and release chlorine and bromine atoms at tropical latitudes.

Large volcanic eruptions have been shown to have substantial albeit uneven ozone-depleting effects, as observed (for example) with the 1991 eruption of Mt. Pinotubo in the Philippines.

Ozone depletion also explains much of the observed reduction in stratospheric and upper tropospheric temperatures. The source of the warmth of the stratosphere is the absorption of UV radiation by ozone, hence reduced ozone leads to cooling. Some stratospheric cooling is also predicted from increases in greenhouse gases such as CO2 and CFCs themselves; however the ozone-induced cooling appears to be dominant.

Predictions of ozone levels remain difficult, but the precision of models' predictions of observed values and the agreement among different modeling techniques have increased steadily. The World Meteorological Organization Global Ozone Research and Monitoring Project—Report No. 44 comes out strongly in favor of the Montreal Protocol, but notes that a UNEP 1994 Assessment overestimated ozone loss for the 1994–1997 period.

Chemicals in the Atmosphere

CFCs and Related Compounds in the Atmosphere

Chlorofluorocarbons (CFCs) and other halogenated ozone depleting substances (ODS) are mainly responsible for man-made chemical ozone depletion. The total amount of effective halogens (chlorine and bromine) in the stratosphere can be calculated and are known as the equivalent effective stratospheric chlorine (EESC).

CFCs were invented by Thomas Midgley, Jr. in the 1920s. They were used in air conditioning and cooling units, as aerosol spray propellants prior to the 1970s, and in the cleaning processes of delicate electronic equipment. They also occur as by-products of some chemical processes. No significant natural sources have ever been identified for these compounds—their presence in the atmosphere is due almost entirely to human manufacture. As mentioned above, when such ozone-depleting chemicals reach the stratosphere, they are dissociated by ultraviolet light to release chlorine atoms. The chlorine atoms act as a catalyst, and each can break down tens of thousands of ozone molecules before being removed from the stratosphere. Given the longevity of CFC molecules, recovery times are measured in decades. It is calculated that a CFC molecule takes an average of about five to seven years to go from the ground level up to the upper atmosphere, and it can stay there for about a century, destroying up to one hundred thousand ozone molecules during that time.

1,1,1-Trichloro-2,2,2-trifluoroethane, also known as CFC-113a, is one of four man-made chemicals newly discovered in the atmosphere by a team at the University of East Anglia. CFC-113a is the only known CFC whose abundance in the atmosphere is still growing. Its source remains a mystery, but illegal manufacturing is suspected by some. CFC-113a seems to have been accumulating unabated since 1960. Between 2010 and 2012, emissions of the gas jumped by 45 percent.

Computer Modeling

Scientists have been increasingly able to attribute the observed ozone depletion to the increase of man-made (anthropogenic) halogen compounds from CFCs by the use of complex chemistry transport models and their validation against observational data (e.g. SLIMCAT, CLaMS—Chemical Lagrangian

Model of the Stratosphere). These models work by combining satellite measurements of chemical concentrations and meteorological fields with chemical reaction rate constants obtained in lab experiments. They are able to identify not only the key chemical reactions but also the transport processes that bring CFC photolysis products into contact with ozone.

Ozone Hole and its Causes

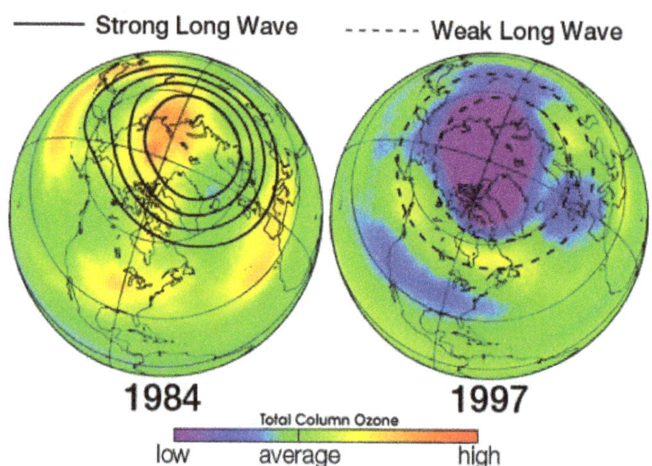

Ozone hole in North America during 1984 (abnormally warm reducing ozone depletion) and 1997 (abnormally cold resulting in increased seasonal depletion). Source: NASA

The Antarctic ozone hole is an area of the Antarctic stratosphere in which the recent ozone levels have dropped to as low as 33 percent of their pre-1975 values. The ozone hole occurs during the Antarctic spring, from September to early December, as strong westerly winds start to circulate around the continent and create an atmospheric container. Within this polar vortex, over 50 percent of the lower stratospheric ozone is destroyed during the Antarctic spring.

As explained above, the primary cause of ozone depletion is the presence of chlorine-containing source gases (primarily CFCs and related halocarbons). In the presence of UV light, these gases dissociate, releasing chlorine atoms, which then go on to catalyze ozone destruction. The Cl-catalyzed ozone depletion can take place in the gas phase, but it is dramatically enhanced in the presence of polar stratospheric clouds (PSCs).

These polar stratospheric clouds (PSC) form during winter, in the extreme cold. Polar winters are dark, consisting of 3 months without solar radiation (sunlight). The lack of sunlight contributes to a decrease in temperature and the polar vortex traps and chills air. Temperatures hover around or below −80 °C. These low temperatures form cloud particles. There are three types of PSC clouds—nitric acid trihydrate clouds, slowly cooling water-ice clouds, and rapid cooling water-ice (nacerous) clouds—provide surfaces for chemical reactions whose products will, in the spring lead to ozone destruction.

The photochemical processes involved are complex but well understood. The key observation is that, ordinarily, most of the chlorine in the stratosphere resides in "reservoir" compounds, primarily chlorine nitrate ($ClONO_2$) as well as stable end products such as HCl. The formation of end products essentially remove Cl from the ozone depletion process. The former sequester Cl, which can be later made available via absorption of light at shorter wavelengths than 400 nm. During the Antarctic winter and spring, however, reactions on the surface of the polar stratospheric cloud particles convert these "reservoir" compounds into reactive free radicals (Cl and ClO). The process by which the clouds remove NO2 from the stratosphere by converting it to nitric acid in the PSC particles, which then are lost by sedimentation is called denitrification. This prevents newly formed ClO from being converted back into ClONO2.

The role of sunlight in ozone depletion is the reason why the Antarctic ozone depletion is greatest during spring. During winter, even though PSCs are at their most abundant, there is no light over the pole to drive chemical reactions. During the spring, however, the sun comes out, providing energy to drive photochemical reactions and melt the polar stratospheric clouds, releasing considerable ClO, which drives the hole mechanism. Further warming temperatures near the end of spring break up the vortex around mid-December. As warm, ozone and NO 2-rich air flows in from lower latitudes, the PSCs are destroyed, the enhanced ozone depletion process shuts down, and the ozone hole closes.

Most of the ozone that is destroyed is in the lower stratosphere, in contrast to the much smaller ozone depletion through homogeneous gas phase reactions, which occurs primarily in the upper stratosphere.

Interest in Ozone Layer Depletion

Public misconceptions and misunderstandings of complex issues like the ozone depletion are common. The limited scientific knowledge of the public led to a confusion with global warming or the perception of global warming as a subset of the "ozone hole". In the beginning, classical green NGOs refrained from using CFC depletion for campaigning, as they assumed the topic was too complicated. They became active much later, e.g. in Greenpeace's support for a CFC-free fridge produced by the former East German company VEB dkk Scharfenstein.

The metaphors used in the CFC discussion (ozone shield, ozone hole) are not "exact" in the scientific sense. The "ozone hole" is more of a *depression*, less "a hole in the windshield". The ozone does not disappear through the layer, nor is there a uniform "thinning" of the ozone layer. However they resonated better with non-scientists and their concerns. The ozone hole was seen as a "hot issue" and imminent risk as lay people feared severe personal consequences such skin cancer, cataracts, damage to plants, and reduction of plankton populations in the ocean's photic zone. Not only on the policy level, ozone regulation compared to climate change fared much better in public opinion. Americans voluntarily switched away from aerosol sprays before legislation was enforced, while climate change failed to achieve comparable concern and public action. The sudden recognition in 1985 that there was a substantial "hole" was widely reported in the press. The especially rapid ozone depletion in Antarctica had previously been dismissed as a measurement error. Scientific consensus was established after regulation.

While the Antarctic ozone hole has a relatively small effect on global ozone, the hole has generated a great deal of public interest because:

- Many have worried that ozone holes might start appearing over other areas of the globe, though to date the only other large-scale depletion is a smaller ozone "dimple" observed during the Arctic spring around the North Pole. Ozone at middle latitudes has declined, but by a much smaller extent (about 4–5 percent decrease).

- If stratospheric conditions become more severe (cooler temperatures, more clouds, more active chlorine), global ozone may decrease at a greater pace. Standard global warming theory predicts that the stratosphere will cool.
- When the Antarctic ozone hole breaks up each year, the ozone-depleted air drifts out into nearby regions. Decreases in the ozone level of up to 10 percent have been reported in New Zealand in the month following the breakup of the Antarctic ozone hole, with ultraviolet-B radiation intensities increasing by more than 15 percent since the 1970s.

Consequences of Ozone Layer Depletion

Since the ozone layer absorbs UVB ultraviolet light from the sun, ozone layer depletion increases surface UVB levels (all else equal), which could lead to damage, including increase in skin cancer. This was the reason for the Montreal Protocol. Although decreases in stratospheric ozone are well-tied to CFCs and to increases in surface UVB, there is no direct observational evidence linking ozone depletion to higher incidence of skin cancer and eye damage in human beings. This is partly because UVA, which has also been implicated in some forms of skin cancer, is not absorbed by ozone, and because it is nearly impossible to control statistics for lifestyle changes over time.

Increased UV

Ozone, while a minority constituent in Earth's atmosphere, is responsible for most of the absorption of UVB radiation. The amount of UVB radiation that penetrates through the ozone layer decreases exponentially with the slant-path thickness and density of the layer. When stratospheric ozone levels decrease, higher levels of UVB reach the Earth's surface. UV-driven phenolic formation in tree rings has dated the start of ozone depletion in northern latitudes to the late 1700s.

In October 2008, the Ecuadorian Space Agency published a report called HIPERION, a study of the last 28 years data from 10 satellites and dozens of ground instruments around the world among them their own, and found that the UV radiation reaching equatorial latitudes was far greater than expected, with the UV Index climbing as high as 24 in some very populated cities; the WHO considers 11 as an extreme index and a great risk to health. The report concluded that depleted ozone levels around the mid-latitudes of the planet are already endangering large populations in these areas. Later, the CONIDA, the Peruvian Space Agency, published its own study, which yielded almost the same findings as the Ecuadorian study.

Biological Effects

The main public concern regarding the ozone hole has been the effects of increased surface UV radiation on human health. So far, ozone depletion in most locations has been typically a few percent and, as noted above, no direct evidence of health damage is available in most latitudes. If the high levels of depletion seen in the ozone hole were to be common across the globe, the effects could be substantially more dramatic. As the ozone hole over Antarctica has in some instances grown so large as to affect parts of Australia, New Zealand, Chile, Argentina, and South Africa, environmentalists have been concerned that the increase in surface UV could be significant.

Ozone depletion would magnify all of the effects of UV on human health, both positive (including production of Vitamin D) and negative (including sunburn, skin cancer, and cataracts). In addition, increased surface UV leads to increased tropospheric ozone, which is a health risk to humans.

Basal and Squamous Cell Carcinomas

The most common forms of skin cancer in humans, basal and squamous cell carcinomas, have been strongly linked to UVB exposure. The mechanism by which UVB induces these cancers is well understood—absorption of UVB radiation causes the pyrimidine bases in the DNA molecule to form dimers, resulting in transcription errors when the DNA replicates. These cancers are relatively mild and rarely fatal, although the treatment of squamous cell carcinoma sometimes requires extensive reconstructive surgery. By combining epidemiological data with results of animal studies, scientists have estimated that every one percent decrease in long-term stratospheric ozone would increase the incidence of these cancers by two percent.

Malignant Melanoma

Another form of skin cancer, malignant melanoma, is much less common but far more dangerous, being lethal in about 15–20 percent of the cases diagnosed. The relationship between malignant melanoma and ultraviolet exposure is not yet fully understood, but it appears that both UVB and UVA are involved. Because of this uncertainty, it is difficult to estimate the effect of ozone depletion on melanoma incidence. One study showed that a 10 percent increase in UVB radiation was associated with a 19 percent increase in melanomas for men and 16 percent for women. A study of people in Punta Arenas, at the southern tip of Chile, showed a 56 percent increase in melanoma and a 46 percent increase in nonmelanoma skin cancer over a period of seven years, along with decreased ozone and increased UVB levels.

Cortical Cataracts

Epidemiological studies suggest an association between ocular cortical cataracts and UVB exposure, using crude approximations of exposure and various cataract assessment techniques. A detailed assessment of ocular exposure to UVB was carried out in a study on Chesapeake Bay Watermen, where increases in average annual ocular exposure were associated with increasing risk of cortical opacity. In this highly exposed group of predominantly white males, the evidence linking cortical opacities to sunlight exposure was the strongest to date. Based on these results, ozone depletion is predicted to cause hundreds of thousands of additional cataracts by 2050.

Increased Tropospheric Ozone

Increased surface UV leads to increased tropospheric ozone. Ground-level ozone is generally recognized to be a health risk, as ozone is toxic due to its strong oxidant properties. The risks are particularly high for young children, the elderly, and those with asthma or other respiratory difficulties. At this time, ozone at ground level is produced mainly by the action of UV radiation on combustion gases from vehicle exhausts.

Increased Production of Vitamin D

Vitamin D is produced in the skin by ultraviolet light. Thus, higher UVB exposure raises human vitamin D in those deficient in it. Recent research (primarily since the Montreal Protocol) shows that many humans have less than optimal vitamin D levels. In particular, in the U.S. population, the lowest quarter of vitamin D (<17.8 ng/ml) were found using information from the National Health and Nutrition Examination Survey to be associated with an increase in all-cause mortality in the general population. While blood level of Vitamin D in excess of 100 ng/ml appear to raise blood calcium excessively and to be associated with higher mortality, the body has mechanisms that prevent sunlight from producing Vitamin D in excess of the body's requirements.

Effects on Non-human Animals

A November 2010 report by scientists at the Institute of Zoology in London found that whales off the coast of California have shown a sharp rise in sun damage, and these scientists "fear that the thinning ozone layer is to blame". The study photographed and took skin biopsies from over 150 whales in the Gulf of California and found "widespread evidence of epidermal damage commonly associated with acute and severe sunburn", having cells that form when the DNA is damaged by UV radiation. The findings suggest "rising UV levels as a result of ozone depletion are to blame for the observed skin damage, in the same way that human skin cancer rates have been on the increase in recent decades."

Effects on Crops

An increase of UV radiation would be expected to affect crops. A number of economically important species of plants, such as rice, depend on cyanobacteria residing on their roots for the retention of nitrogen. Cyanobacteria are sensitive to UV radiation and would be affected by its increase. "Despite mechanisms to reduce or repair the effects of increased ultraviolet radiation, plants have a limited ability to adapt to increased levels of UVB, therefore plant growth can be directly affected by UVB radiation."

Public Policy

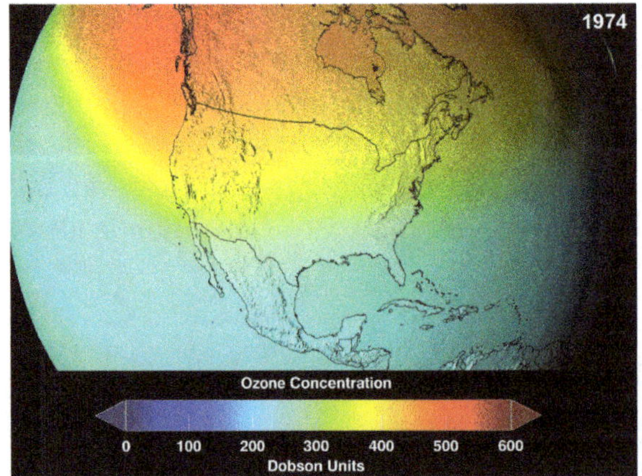

NASA projections of stratospheric ozone concentrations if chlorofluorocarbons had not been banned

The full extent of the damage that CFCs have caused to the ozone layer is not known and will not be known for decades; however, marked decreases in column ozone have already been observed. The

Montreal and Vienna conventions were installed long before a scientific consensus was established or important uncertainties in the science field were being resolved. The ozone case was understood comparably well by lay persons as e.g. *Ozone shield* or *ozone hole* were useful "easy-to-understand bridging metaphors". Americans voluntarily switched away from aerosol sprays, resulting in a 50 percent sales loss even before legislation was enforced.

After a 1976 report by the United States National Academy of Sciences concluded that credible scientific evidence supported the ozone depletion hypothesis a few countries, including the United States, Canada, Sweden, Denmark, and Norway, moved to eliminate the use of CFCs in aerosol spray cans. At the time this was widely regarded as a first step towards a more comprehensive regulation policy, but progress in this direction slowed in subsequent years, due to a combination of political factors (continued resistance from the halocarbon industry and a general change in attitude towards environmental regulation during the first two years of the Reagan administration) and scientific developments (subsequent National Academy assessments that indicated that the first estimates of the magnitude of ozone depletion had been overly large). A critical DuPont manufacturing patent for Freon was set to expire in 1979. The United States banned the use of CFCs in aerosol cans in 1978. The European Community rejected proposals to ban CFCs in aerosol sprays, and in the U.S., CFCs continued to be used as refrigerants and for cleaning circuit boards. Worldwide CFC production fell sharply after the U.S. aerosol ban, but by 1986 had returned nearly to its 1976 level. In 1993, DuPont shut down its CFC facility.

The U.S. Government's attitude began to change again in 1983, when William Ruckelshaus replaced Anne M. Burford as Administrator of the United States Environmental Protection Agency. Under Ruckelshaus and his successor, Lee Thomas, the EPA pushed for an international approach to halocarbon regulations. In 1985 20 nations, including most of the major CFC producers, signed the Vienna Convention for the Protection of the Ozone Layer, which established a framework for negotiating international regulations on ozone-depleting substances. That same year, the discovery of the Antarctic ozone hole was announced, causing a revival in public attention to the issue. In 1987, representatives from 43 nations signed the Montreal Protocol. Meanwhile, the halocarbon industry shifted its position and started supporting a protocol to limit CFC production. However, this shift was uneven with DuPont acting more quickly than their European counterparts. DuPont may have feared court action related to increased skin cancer especially as the EPA had published a study in 1986 claiming that an additional 40 million cases and 800,000 cancer deaths were to be expected in the U.S. in the next 88 years. The EU shifted its position as well after Germany gave up its defence of the CFC industry and started supporting moves towards regulation. Government and industry in France and the UK tried to defend their CFC producing industries even after the Montreal Protocol had been signed.

At Montreal, the participants agreed to freeze production of CFCs at 1986 levels and to reduce production by 50 percent by 1999. After a series of scientific expeditions to the Antarctic produced convincing evidence that the ozone hole was indeed caused by chlorine and bromine from manmade organohalogens, the Montreal Protocol was strengthened at a 1990 meeting in London. The participants agreed to phase out CFCs and halons entirely (aside from a very small amount marked for certain "essential" uses, such as asthma inhalers) by 2000 in non-Article 5 countries and by 2010 in Article 5 (less developed) signatories. At a 1992 meeting in Copenhagen, the phase-out date was moved up to 1996. At the same meeting, methyl bromide (MeBr), a fumigant used primarily in agricultural production, was added to the list of controlled substances. For all substances controlled under the protocol, phaseout schedules were delayed for less developed ('Article 5(1)') countries, and phaseout in these countries was supported by transfers of expertise, technology, and money from non-Article 5(1) Parties to the Protocol. Additionally, exemptions from the agreed schedules could be applied for under the Essential Use Exemption (EUE) process for substances other than methyl bromide and under the Critical Use Exemption (CUE) process for methyl bromide.

A hydrocarbon refrigerant was developed in 1992 at the request of the non-governmental organization (NGO) Greenpeace, and was being used by some 40 percent of the refrigerator market in 2013. In the U.S., however, change has been much slower. To some extent, CFCs were being replaced by the less damaging hydrochlorofluorocarbons (HCFCs), although concerns remain regarding HCFCs also. In some applications, hydrofluorocarbons (HFCs) were being used to replace CFCs. HFCs, which contain no chlorine or bromine, do not contribute at all to ozone depletion although they are potent greenhouse gases. The best known of these compounds is probably HFC-134a (R-134a), which in the United States has largely replaced CFC-12 (R-12) in automobile air conditioners. In laboratory analytics (a former "essential" use) the ozone depleting substances can be replaced with various other solvents. The development and promotion of an ozone-safe hydrocarbon refrigerant was a breakthrough which occurred after the initiative of a non-governmental organization (NGO). Civil society including especially NGOs, in fact, played critical roles at all stages of policy development leading up to the Vienna Conference, the Montreal Protocol, and in assessing compliance afterwards. The major companies claimed that no alternatives to HFC existed. By 1992, Greenpeace had requested a scientific team to research an ozone-safe refrigerant, an effort which resulted in a successful hydrocarbon formula, for which Greenpeace left the patent as open source. The NGO maintained the technology as an open patent as it worked successfully first with a small company to market an appliance beginning in Europe, then Asia and later Latin America, receiving a 1997 UNEP award. By 1995, Germany had already made CFC refrigerators illegal. Production spread to companies like Electrolux, Bosch, and LG, with sales reaching some 300 million refrigerators by 2008. However, the giant corporations all continued to refuse to apply the technology in Latin America. In 2003, a domestic Argentinian company began Greenfreeze production, while the giant Bosch in Brazil began a year later. Chemical companies like Du Pont, whose representatives even disparaged the green technology as "that German technology," maneuvered the EPA to block the technology in the U.S. until 2011. Ben & Jerry's of Unilever and General Electric, spurred by Greenpeace, had expressed formal interest in 2008 which figured in the EPA's final approval. Currently, more than 600 million refrigerators have been sold, making up more than 40 percent of the market. Since 2004, corporations like Coca-Cola, Carlsberg, and IKEA have been forming a coalition to promote the ozone-safe Greenfreeze units.

More recently, policy experts have advocated for efforts to link ozone protection efforts to climate protection efforts. Many ODS are also greenhouse gases, some thousands of times more powerful agents of radiative forcing than carbon dioxide over the short and medium term. Thus policies protecting the ozone layer have had benefits in mitigating climate change. In fact, the reduction of the radiative forcing due to ODS probably masked the true level of climate change effects of other GHGs, and was responsible for the "slow down" of global warming from the mid-90s. Policy decisions in one arena affect the costs and effectiveness of environmental improvements in the other.

ODS Requirements in the Marine Industry

The IMO has amended MARPOL Annex VI Regulation 12 regarding ozone depleting substances. As from July 1, 2010, all vessels where MARPOL Annex VI is applicable should have a list of equipment using ozone depleting substances. The list should include name of ODS, type and location of equipment, quantity in kg and date. All changes since that date should be recorded in an ODS Record book on board recording all intended or unintended releases to the atmosphere. Furthermore, new ODS supply or landing to shore facilities should be recorded as well.

Prospects of Ozone Depletion

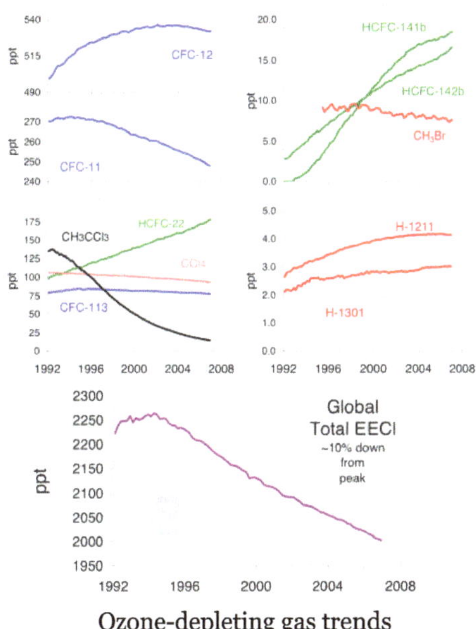

Ozone-depleting gas trends

Since the adoption and strengthening of the Montreal Protocol has led to reductions in the emissions of CFCs, atmospheric concentrations of the most-significant compounds have been declining. These substances are being gradually removed from the atmosphere; since peaking in 1994, the Effective Equivalent Chlorine (EECl) level in the atmosphere had dropped about 10 percent by 2008. The decrease in ozone-depleting chemicals has also been significantly affected by a decrease in bromine-containing chemicals. The data suggest that substantial natural sources exist for atmospheric methyl bromide (CH3Br). The phase-out of CFCs means that nitrous oxide (N2O), which is not covered by the Montreal Protocol, has become the most highly emitted ozone-depleting substance and is expected to remain so throughout the 21st century.

A 2005 IPCC review of ozone observations and model calculations concluded that the global amount of ozone has now approximately stabilized. Although considerable variability is expected from year to year, including in polar regions where depletion is largest, the ozone layer is expected to begin to recover in coming decades due to declining ozone-depleting substance concentrations, assuming full compliance with the Montreal Protocol.

The Antarctic ozone hole is expected to continue for decades. Ozone concentrations in the lower stratosphere over Antarctica will increase by 5–10 percent by 2020 and return to pre-1980 levels by about 2060–2075. This is 10–25 years later than predicted in earlier assessments, because of revised estimates of atmospheric concentrations of ozone-depleting substances, including a larger predicted future usage in developing countries. Another factor that may prolong ozone depletion is the drawdown of nitrogen oxides from above the stratosphere due to changing wind patterns. A gradual trend toward "healing" was reported in 2016.

Research History

The basic physical and chemical processes that lead to the formation of an ozone layer in the Earth's stratosphere were discovered by Sydney Chapman in 1930. Short-wavelength UV radiation splits an oxygen (O2) molecule into two oxygen (O) atoms, which then combine with other oxygen molecules to form ozone. Ozone is removed when an oxygen atom and an ozone molecule "recombine" to form two oxygen molecules, i.e. $O + O3 \rightarrow 2O2$. In the 1950s, David Bates and Marcel Nicolet presented evidence that various free radicals, in particular hydroxyl (OH) and nitric oxide (NO), could catalyze this recombination reaction, reducing the overall amount of ozone. These free radicals were known to be present in the stratosphere, and so were regarded as part of the natural balance—it was estimated that in their absence, the ozone layer would be about twice as thick as it currently is.

In 1970 Paul Crutzen pointed out that emissions of nitrous oxide (N2O), a stable, long-lived gas produced by soil bacteria, from the Earth's surface could affect the amount of nitric oxide (NO) in the stratosphere. Crutzen showed that nitrous oxide lives long enough to reach the stratosphere, where it is converted into NO. Crutzen then noted that increasing use of fertilizers might have led to an increase in nitrous oxide emissions over the natural background, which would in turn result in an increase in the amount of NO in the stratosphere. Thus human activity could affect the stratospheric ozone layer. In the following year, Crutzen and (independently) Harold Johnston suggested that NO emissions from supersonic passenger aircraft, which would fly in the lower stratosphere, could also deplete the ozone layer. However, more recent analysis in 1995 by David W. Fahey, an atmospheric scientist at the National Oceanic and Atmospheric Administration, found that the drop in ozone would be from 1–2 percent if a fleet of 500 supersonic passenger aircraft were operated. This, Fahey expressed, would not be a showstopper for advanced supersonic passenger aircraft development.

Rowland–molina Hypothesis

In 1974 Frank Sherwood Rowland, Chemistry Professor at the University of California at Irvine, and his postdoctoral associate Mario J. Molina suggested that long-lived organic halogen compounds, such as CFCs, might behave in a similar fashion as Crutzen had proposed for nitrous oxide. James Lovelock had recently discovered, during a cruise in the South Atlantic in 1971, that almost all of the CFC compounds manufactured since their invention in 1930 were still present in the atmosphere. Molina and Rowland concluded that, like N2O, the CFCs would reach the stratosphere where they would be dissociated by UV light, releasing chlorine atoms. A year earlier, Richard Stolarski and Ralph Cicerone at the University of Michigan had shown that Cl is even more efficient than NO at catalyzing the destruction of ozone. Similar conclusions were reached by Michael McElroy and Steven Wofsy at Harvard University. Neither group, however, had realized that CFCs were a potentially large source of stratospheric chlorine—instead, they had been investigating the possible effects of HCl emissions from the Space Shuttle, which are very much smaller.

The Rowland–Molina hypothesis was strongly disputed by representatives of the aerosol and halocarbon industries. The Chair of the Board of DuPont was quoted as saying that ozone depletion theory is "a science fiction tale ... a load of rubbish ... utter nonsense". Robert Abplanalp, the President of Precision Valve Corporation (and inventor of the first practical aerosol spray can valve), wrote to the Chancellor of UC Irvine to complain about Rowland's public statements. Nevertheless, within three years most of the basic assumptions made by Rowland and Molina were confirmed by laboratory measurements and by direct observation in the stratosphere. The concentrations of the source gases (CFCs and related compounds) and the chlorine reservoir species (HCl and ClO-NO2) were measured throughout the stratosphere, and demonstrated that CFCs were indeed the major source of stratospheric chlorine, and that nearly all of the CFCs emitted would eventually reach the stratosphere. Even more convincing was the measurement, by James G. Anderson and collaborators, of chlorine monoxide (ClO) in the stratosphere. ClO is produced by the reaction of Cl with ozone—its observation thus demonstrated that Cl radicals not only were present in the stratosphere but also were actually involved in destroying ozone. McElroy and Wofsy extended the work of Rowland and Molina by showing that bromine atoms were even more effective catalysts for ozone loss than chlorine atoms and argued that the brominated organic compounds known as halons, widely used in fire extinguishers, were a potentially large source of stratospheric bromine. In 1976 the United States National Academy of Sciences released a report concluding that the ozone depletion hypothesis was strongly supported by the scientific evidence. Scientists calculated that if CFC production continued to increase at the going rate of 10 percent per year until 1990 and then remain steady, CFCs would cause a global ozone loss of 5–7 percent by 1995, and a 30–50 percent loss by 2050. In response the United States, Canada and Norway banned the use of CFCs in aerosol spray cans in 1978. However, subsequent research, summarized by the National Academy in reports issued between 1979 and 1984, appeared to show that the earlier estimates of global ozone loss had been too large.

Crutzen, Molina, and Rowland were awarded the 1995 Nobel Prize in Chemistry for their work on stratospheric ozone.

Antarctic Ozone Hole

The discovery of the Antarctic "ozone hole" by British Antarctic Survey scientists Farman, Gardiner and Shanklin (first reported in a paper in *Nature* in May 1985) came as a shock to the scientific community, because the observed decline in polar ozone was far larger than anyone had anticipated. Satellite measurements showing massive depletion of ozone around the south pole were becoming available at the same time. However, these were initially rejected as unreasonable by data quality control algorithms (they were filtered out as errors since the values were unexpectedly low); the ozone hole was detected only in satellite data when the raw data was reprocessed following evidence of ozone depletion in *in situ* observations. When the software was rerun without the flags, the ozone hole was seen as far back as 1976.

Susan Solomon, an atmospheric chemist at the National Oceanic and Atmospheric Administration (NOAA), proposed that chemical reactions on polar stratospheric clouds (PSCs) in the cold Antarctic stratosphere caused a massive, though localized and seasonal, increase in the amount of chlorine present in active, ozone-destroying forms. The polar stratospheric clouds in Antarctica are only formed when there are very low temperatures, as low as −80 °C, and early spring conditions.

In such conditions the ice crystals of the cloud provide a suitable surface for conversion of unreactive chlorine compounds into reactive chlorine compounds, which can deplete ozone easily.

Moreover, the polar vortex formed over Antarctica is very tight and the reaction occurring on the surface of the cloud crystals is far different from when it occurs in atmosphere. These conditions have led to ozone hole formation in Antarctica. This hypothesis was decisively confirmed, first by laboratory measurements and subsequently by direct measurements, from the ground and from high-altitude airplanes, of very high concentrations of chlorine monoxide (ClO) in the Antarctic stratosphere.

Alternative hypotheses, which had attributed the ozone hole to variations in solar UV radiation or to changes in atmospheric circulation patterns, were also tested and shown to be untenable.

Meanwhile, analysis of ozone measurements from the worldwide network of ground-based Dobson spectrophotometers led an international panel to conclude that the ozone layer was in fact being depleted, at all latitudes outside of the tropics. These trends were confirmed by satellite measurements. As a consequence, the major halocarbon-producing nations agreed to phase out production of CFCs, halons, and related compounds, a process that was completed in 1996.

Since 1981 the United Nations Environment Programme, under the auspices of the World Meteorological Organization, has sponsored a series of technical reports on the Scientific Assessment of Ozone Depletion, based on satellite measurements. The 2007 report showed that the hole in the ozone layer was recovering and the smallest it had been for about a decade. The 2010 report found, "Over the past decade, global ozone and ozone in the Arctic and Antarctic regions is no longer decreasing but is not yet increasing. The ozone layer outside the Polar regions is projected to recover to its pre-1980 levels some time before the middle of this century. In contrast, the springtime ozone hole over the Antarctic is expected to recover much later." In 2012, NOAA and NASA reported "Warmer air temperatures high above the Antarctic led to the second smallest season ozone hole in 20 years averaging 17.9 million square kilometres. The hole reached its maximum size for the season on Sept 22, stretching to 21.2 million square kilometres." A gradual trend toward "healing" was reported in 2016.

The hole in the Earth's ozone layer over the South Pole has affected atmospheric circulation in the Southern Hemisphere all the way to the equator. The ozone hole has influenced atmospheric circulation all the way to the tropics and increased rainfall at low, subtropical latitudes in the Southern Hemisphere.

Arctic Ozone Hole

On March 15, 2011, a record ozone layer loss was observed, with about half of the ozone present over the Arctic having been destroyed. The change was attributed to increasingly cold winters in the Arctic stratosphere at an altitude of approximately 20 km (12 mi), a change associated with global warming in a relationship that is still under investigation. By March 25, the ozone loss had become the largest compared to that observed in all previous winters with the possibility that it would become an ozone hole. This would require that the quantities of ozone to fall below 200 Dobson units, from the 250 recorded over central Siberia. It is predicted that the thinning layer would affect parts of Scandinavia and Eastern Europe on March 30–31.

On March 3, 2005, the journal Nature published an article linking 2004's unusually large Arctic ozone hole to solar wind activity.

On October 2, 2011, a study was published in the journal *Nature*, which said that between December 2010 and March 2011 up to 80 percent of the ozone in the atmosphere at about 20 kilometres (12 mi) above the surface was destroyed. The level of ozone depletion was severe enough that scientists said it could be compared to the ozone hole that forms over Antarctica every winter. According to the study, "for the first time, sufficient loss occurred to reasonably be described as an Arctic ozone hole." The study analyzed data from the Aura and CALIPSO satellites, and determined that the larger-than-normal ozone loss was due to an unusually long period of cold weather in the Arctic, some 30 days more than typical, which allowed for more ozone-destroying chlorine compounds to be created. According to Lamont Poole, a co-author of the study, cloud and aerosol particles on which the chlorine compounds are found "were abundant in the Arctic until mid March 2011—much later than usual—with average amounts at some altitudes similar to those observed in the Antarctic, and dramatically larger than the near-zero values seen in March in most Arctic winters".

Tibet Ozone Hole

As winters that are colder are more affected, at times there is an ozone hole over Tibet. In 2006, a 2.5 million square kilometer ozone hole was detected over Tibet. Also again in 2011 an ozone hole appeared over mountainous regions of Tibet, Xinjiang, Qinghai and the Hindu Kush, along with an unprecedented hole over the Arctic, though the Tibet one is far less intense than the ones over the Arctic or Antarctic.

Potential Depletion by Storm Clouds

Research in 2012 showed that the same process that produces the ozone hole over Antarctica occurs over summer storm clouds in the United States, and thus may be destroying ozone there as well.

Ozone Depletion and Global Warming

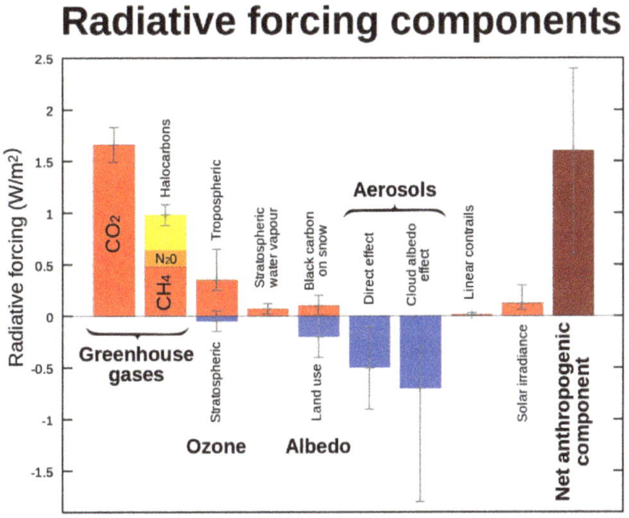

Radiative forcing from various greenhouse gases and other sources

Among others, Robert Watson had a role in the science assessment and in the regulation efforts of ozone depletion and global warming. Prior to the 1980s, the EU, NASA, NAS, UNEP, WMO and the British government had dissenting scientific reports and Watson played a crucial role in the process of unified assessments. Based on the experience with the ozone case, the IPCC started to work on a unified reporting and science assessment to reach a consensus to provide the IPCC Summary for Policymakers.

There are various areas of linkage between ozone depletion and global warming science:

- The same CO2 radiative forcing that produces global warming is expected to cool the stratosphere. This cooling, in turn, is expected to produce a relative *increase* in ozone (O3) depletion in polar area and the frequency of ozone holes.
- Conversely, ozone depletion represents a radiative forcing of the climate system. There are two opposing effects: Reduced ozone causes the stratosphere to absorb less solar radiation, thus cooling the stratosphere while warming the troposphere; the resulting colder stratosphere emits less long-wave radiation downward, thus cooling the troposphere. Overall, the cooling dominates; the IPCC concludes "*observed stratospheric O3 losses over the past two decades have caused a negative forcing of the surface-troposphere system*" of about -0.15 ± 0.10 watts per square meter (W/m^2).
- One of the strongest predictions of the greenhouse effect is that the stratosphere will cool. Although this cooling has been observed, it is not trivial to separate the effects of changes in the concentration of greenhouse gases and ozone depletion since both will lead to cooling. However, this can be done by numerical stratospheric modeling. Results from the National Oceanic and Atmospheric Administration's Geophysical Fluid Dynamics Laboratory show that above 20 km (12 mi), the greenhouse gases dominate the cooling.
- As noted under 'Public Policy', ozone depleting chemicals are also often greenhouse gases. The increases in concentrations of these chemicals have produced 0.34 ± 0.03 W/m^2 of radiative forcing, corresponding to about 14 percent of the total radiative forcing from increases in the concentrations of well-mixed greenhouse gases.
- The long term modeling of the process, its measurement, study, design of theories and testing take decades to document, gain wide acceptance, and ultimately become the dominant paradigm. Several theories about the destruction of ozone were hypothesized in the 1980s, published in the late 1990s, and are currently being investigated. Dr Drew Schindell, and Dr Paul Newman, NASA Goddard, proposed a theory in the late 1990s, using computational modeling methods to model ozone destruction, that accounted for 78 percent of the ozone destroyed. Further refinement of that model accounted for 89 percent of the ozone destroyed, but pushed back the estimated recovery of the ozone hole from 75 years to 150 years. (An important part of that model is the lack of stratospheric flight due to depletion of fossil fuels.)

Misconceptions

CFC Weight

Since CFC molecules are heavier than air (nitrogen or oxygen), it is commonly believed that the CFC molecules cannot reach the stratosphere in significant amount. However, atmospheric gases are not sorted by weight; the forces of wind can fully mix the gases in the atmosphere. Lighter CFCs are evenly distributed throughout the turbosphere and reach the upper atmosphere, although some of the heavier CFCs are not evenly distributed.

Percentage of Man-made Chlorine

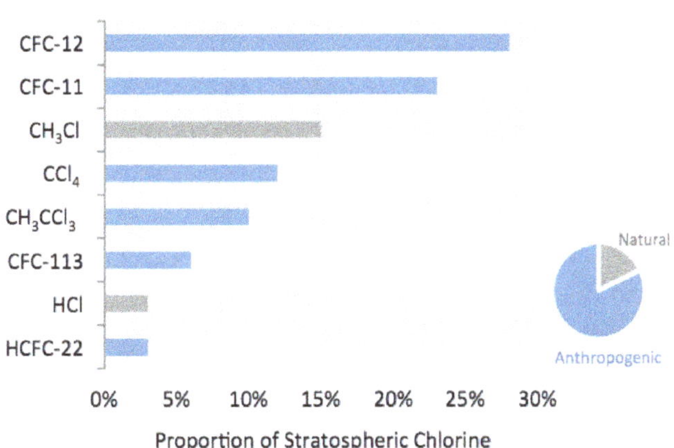

Sources of Stratospheric Chlorine

Another misconception is that "it is generally accepted that natural sources of tropospheric chlorine are four to five times larger than man-made ones." While strictly true, *tropospheric* chlorine is irrelevant; it is *stratospheric* chlorine that affects ozone depletion. Chlorine from ocean spray is soluble and thus is washed by rainfall before it reaches the stratosphere. CFCs, in contrast, are insoluble and long-lived, allowing them to reach the stratosphere. In the lower atmosphere, there is much more chlorine from CFCs and related haloalkanes than there is in HCl from salt spray, and in the stratosphere halocarbons are dominant. Only methyl chloride, which is one of these halocarbons, has a mainly natural source, and it is responsible for about 20 percent of the chlorine in the stratosphere; the remaining 80 percent comes from manmade sources.

Very violent volcanic eruptions can inject HCl into the stratosphere, but researchers have shown that the contribution is not significant compared to that from CFCs. A similar erroneous assertion is that soluble halogen compounds from the volcanic plume of Mount Erebus on Ross Island, Antarctica are a major contributor to the Antarctic ozone hole.

Nevertheless, the latest study showed that the role of Mount Erebus volcano in the Antarctic ozone depletion was probably underestimated. Based on the NCEP/NCAR reanalysis data over the last 35 years and by using the NOAA HYSPLIT trajectory model, researchers showed that Erebus volcano gas emissions (including hydrogen chloride (HCl)) can reach the Antarctic stratosphere via high-latitude cyclones and then the polar vortex. Depending on Erebus volcano activity, the additional annual HCl mass entering the stratosphere from Erebus varies from 1.0 to 14.3 kt.

First Observation

G.M.B. Dobson (Exploring the Atmosphere, 2nd Edition, Oxford, 1968) mentioned that when springtime ozone levels in the Antarctic over Halley Bay were first measured in 1956, he was surprised to find that they were ~320 DU, or about 150 DU below spring Arctic levels of ~450 DU. These were at that time the only known Antarctic ozone values available. What Dobson describes is essentially the *baseline* from which the ozone hole is measured: actual ozone hole values are in the 150–100 DU range.

The discrepancy between the Arctic and Antarctic noted by Dobson was primarily a matter of timing: during the Arctic spring ozone levels rose smoothly, peaking in April, whereas in the Antarctic they stayed approximately constant during early spring, rising abruptly in November when the polar vortex broke down.

The behavior seen in the Antarctic ozone hole is completely different. Instead of staying constant, early springtime ozone levels suddenly drop from their already low winter values, by as much as 50 percent, and normal values are not reached again until December.

Location of Hole

Some people thought that the ozone hole should be above the sources of CFCs. However, CFCs are well mixed globally in the troposphere and stratosphere. The reason for occurrence of the ozone hole above Antarctica is not because there are more CFCs concentrated but because the low temperatures help form polar stratospheric clouds. In fact, there are findings of significant and localized "ozone holes" above other parts of the earth.

World Ozone Day

In 1994, the United Nations General Assembly voted to designate September 16 as "World Ozone Day", to commemorate the signing of the Montreal Protocol on that date in 1987.

References

- Roan, Sharon (1989) Ozone crisis: The 15-year evolution of a sudden global emergency, New York: Wiley, p. 56 ISBN 0-471-52823-4
- "Climate Change 2001: Working Group I: The Scientific Basis". Intergovernmental Panel on Climate Change Work Group I. 2001. pp. Chapter 6.4 Stratospheric Ozone. Retrieved May 28, 2016.
- "Reconstruction of Paleobehavior of Ozonosphere Based on Response to UV-B Radiation Effect in Dendrochronologic Signal" (PDF). Atmospheric Radiation Measurement, USA. Retrieved May 28, 2016.
- "Lanzan la primera de las "Propuestas Greenpeace": la heladera "Greenfreeze" | Greenpeace Argentina". Greenpeace.org. Retrieved September 27, 2015.
- Lipkin, Richard (October 7, 1995). SST emissions cut stratospheric ozone. (The introduction of 500 new supersonic transport planes by 2015 could deplete the ozone layer by as much as 1%). Science News.
- Fountain, Henry (July 27, 2012). "Storms Threaten Ozone Layer Over U.S., Study Says". The New York Times. p. A1. Retrieved March 13, 2015.
- "The Relative Roles of Ozone and Other Greenhouse Gases in Climate Change in the Stratosphere". Geophysical Fluid Dynamics Laboratory. February 29, 2004. Archived from the original on January 20, 2009. Retrieved March 13, 2015. CS1 maint: Unfit url (link)
- Johannes C. Laube; Mike J. Newland; Christopher Hogan; Carl A. M. Brenninkmeijer; Paul J. Fraser; Patricia Martinerie; David E. Oram; Claire E. Reeves; Thomas Röckmann; Jakob Schwander; Emmanuel Witrant; William T. Sturges (March 9, 2014). "Newly detected ozone-depleting substances in the atmosphere". Nature Geoscience. 7 (4): 266–269. doi:10.1038/ngeo2109. Retrieved March 10, 2014.

- Shabecoff, Philip (November 5, 1986). "U.S. Report Predicts Rise in Skin Cancer with Loss of Ozone". The New York Times. p. A1. Retrieved January 10, 2013.
- "Increase in supersonic jets could be threat to ozone U-2 plane trails Concorde, studies exhaust particles". The Baltimore Sun. Newsday. October 8, 1995. Retrieved December 21, 2012.
- "NOAA Study Shows Nitrous Oxide Now Top Ozone-Depleting Emission". Noaanews.noaa.gov. August 27, 2009. Retrieved April 6, 2011.
- "The Arctic Ozone Sieve: More Global Weirding?". Scienceblogs.com. March 25, 2011. Archived from the original on April 4, 2011. Retrieved April 6.
- "Arctic ozone loss at record level". BBC News Online. October 2, 2011. Archived from the original on October 2, 2011. Retrieved October 3, 2011.
- "Earth news: Chinese Scientists Find New Ozone Hole Over Tibet". Elainemeinelsupkis.typepad.com. May 4, 2006. Retrieved April 6, 2011.
- Schiermeier, Quirin (February 22, 1999). "The Great Beyond: Arctic ozone hole causes concern". Blogs.nature.com. Retrieved April 6, 2011.
- "Twenty Questions and Answers About the Ozone Layer". Scientific Assessment of Ozone Depletion: 2010 (PDF). World Meteorological Organization. 2011. Retrieved March 13, 2015.
- "Nash, Eric; Newman, Paul (September 19, 2001). "NASA Confirms Arctic Ozone Depletion Trigger". Image of the Day. NASA. Retrieved April 16, 2011.
- Toon, Owen B.; Turco, Richard P. (June 1991). "Polar Stratospheric Clouds and Ozone Depletion" (PDF). Scientific American. 264 (6): 68–74. Bibcode:1991SciAm.264...68T. doi:10.1038/scientificamerican0691-68. Retrieved April 16, 2011.

6

Air Pollution Analyser

Atmospheric dispersion modeling is the explanation of how air pollutants are dispersed in the atmosphere. It can be used to predict future issues caused because of specific scenarios. Some of the topics explained in this text are outline of air pollution dispersion and roadway air dispersion modeling. This section has been carefully written to provide an easy understanding of air pollution analyzers.

Atmospheric Dispersion Modeling

Atmospheric dispersion modeling is the mathematical simulation of how air pollutants disperse in the ambient atmosphere. It is performed with computer programs that solve the mathematical equations and algorithms which simulate the pollutant dispersion. The dispersion models are used to estimate the downwind ambient concentration of air pollutants or toxins emitted from sources such as industrial plants, vehicular traffic or accidental chemical releases. They can also be used to predict future concentrations under specific scenarios (i.e. changes in emission sources). Therefore, they are the dominant type of model used in air quality policy making. They are most useful for pollutants that are dispersed over large distances and that may react in the atmosphere. For pollutants that have a very high spatio-temporal variability (i.e. have very steep distance to source decay such as black carbon) and for epidemiological studies statistical land-use regression models are also used.

Dispersion models are important to governmental agencies tasked with protecting and managing the ambient air quality. The models are typically employed to determine whether existing or proposed new industrial facilities are or will be in compliance with the National Ambient Air Quality Standards (NAAQS) in the United States and other nations. The models also serve to assist in the design of effective control strategies to reduce emissions of harmful air pollutants. During the late 1960s, the Air Pollution Control Office of the U.S. EPA initiated research projects that would lead to the development of models for the use by urban and transportation planners. A major and significant application of a roadway dispersion model that resulted from such research was applied to the Spadina Expressway of Canada in 1971.

Air dispersion models are also used by public safety responders and emergency management personnel for emergency planning of accidental chemical releases. Models are used to determine the consequences of accidental releases of hazardous or toxic materials, Accidental releases may result in fires, spills or explosions that involve hazardous materials, such as chemicals or radionuclides. The results of dispersion modeling, using worst case accidental release source terms and meteorological conditions, can provide an estimate of location impacted areas, ambient concentrations, and be used to determine protective actions appropriate in the event a release occurs. Appropriate protective actions may include evacuation or shelter in place for persons in the downwind direction. At industrial facilities, this type of consequence assessment or emergency planning is required under the Clean Air Act (United States) (CAA) codified in Part 68 of Title 40 of the Code of Federal Regulations.

The dispersion models vary depending on the mathematics used to develop the model, but all require the input of data that may include:

- Meteorological conditions such as wind speed and direction, the amount of atmospheric turbulence (as characterized by what is called the "stability class"), the ambient air temperature, the height to the bottom of any inversion aloft that may be present, cloud cover and solar radiation.
- Source term (the concentration or quantity of toxins in emission or accidental release source terms) and temperature of the material
- Emissions or release parameters such as source location and height, type of source (i.e., fire, pool or vent stack)and exit velocity, exit temperature and mass flow rate or release rate.
- Terrain elevations at the source location and at the receptor location(s), such as nearby homes, schools, businesses and hospitals.
- The location, height and width of any obstructions (such as buildings or other structures) in the path of the emitted gaseous plume, surface roughness or the use of a more generic parameter "rural" or "city" terrain.

Many of the modern, advanced dispersion modeling programs include a pre-processor module for the input of meteorological and other data, and many also include a post-processor module for graphing the output data and/or plotting the area impacted by the air pollutants on maps. The plots of areas impacted may also include isopleths showing areas of minimal to high concentrations that define areas of the highest health risk. The isopleths plots are useful in determining protective actions for the public and responders.

The atmospheric dispersion models are also known as atmospheric diffusion models, air dispersion models, air quality models, and air pollution dispersion models.

Atmospheric Layers

Discussion of the layers in the Earth's atmosphere is needed to understand where airborne pollutants disperse in the atmosphere. The layer closest to the Earth's surface is known as the *troposphere*. It extends from sea-level to a height of about 18 km and contains about 80 percent of the mass of the overall atmosphere. The *stratosphere* is the next layer and extends from 18 km to about 50 km. The third layer is the *mesosphere* which extends from 50 km to about 80 km. There are other layers above 80 km, but they are insignificant with respect to atmospheric dispersion modeling.

The lowest part of the troposphere is called the *atmospheric boundary layer (ABL)* or the *planetary boundary layer (PBL)* and extends from the Earth's surface to about 1.5 to 2.0 km in height.

The air temperature of the atmospheric boundary layer decreases with increasing altitude until it reaches what is called the *inversion layer* (where the temperature increases with increasing altitude) that caps the atmospheric boundary layer. The upper part of the troposphere (i.e., above the inversion layer) is called the *free troposphere* and it extends up to the 18 km height of the troposphere.

The ABL is of the most important with respect to the emission, transport and dispersion of airborne pollutants. The part of the ABL between the Earth's surface and the bottom of the inversion layer is known as the mixing layer. Almost all of the airborne pollutants emitted into the ambient atmosphere are transported and dispersed within the mixing layer. Some of the emissions penetrate the inversion layer and enter the free troposphere above the ABL.

In summary, the layers of the Earth's atmosphere from the surface of the ground upwards are: the ABL made up of the mixing layer capped by the inversion layer; the free troposphere; the stratosphere; the mesosphere and others. Many atmospheric dispersion models are referred to as *boundary layer models* because they mainly model air pollutant dispersion within the ABL. To avoid confusion, models referred to as *mesoscale models* have dispersion modeling capabilities that extend horizontally up to a few hundred kilometres. It does not mean that they model dispersion in the mesosphere.

Gaussian Air Pollutant Dispersion Equation

The technical literature on air pollution dispersion is quite extensive and dates back to the 1930s and earlier. One of the early air pollutant plume dispersion equations was derived by Bosanquet and Pearson. Their equation did not assume Gaussian distribution nor did it include the effect of ground reflection of the pollutant plume.

Sir Graham Sutton derived an air pollutant plume dispersion equation in 1947 which did include the assumption of Gaussian distribution for the vertical and crosswind dispersion of the plume and also included the effect of ground reflection of the plume.

Under the stimulus provided by the advent of stringent environmental control regulations, there was an immense growth in the use of air pollutant plume dispersion calculations between the late 1960s and today. A great many computer programs for calculating the dispersion of air pollutant emissions were developed during that period of time and they were called "air dispersion models". The basis for most of those models was the Complete Equation For Gaussian Dispersion Modeling Of Continuous, Buoyant Air Pollution Plumes shown below:

$$C = \frac{Q}{u} \cdot \frac{f}{\sigma_y \sqrt{2\pi}} \cdot \frac{g_1 + g_2 + g_3}{\sigma_z \sqrt{2\pi}}$$

where:

f = crosswind dispersion parameter

= $\exp[-y^2/(2\sigma_y^2)]$

g = vertical dispersion parameter = $g_1 + g_2 + g_3$

g_1 = vertical dispersion with no reflections

$= \exp[-(z-H)^2/(2\sigma_z^2)]$

g_2 = vertical dispersion for reflection from the ground

$= \exp[-(z+H)^2/(2\sigma_z^2)]$

g_3 = vertical dispersion for reflection from an inversion aloft

$$= \sum_{m=1}^{\infty} \{\exp[-(z-H-2mL)^2/(2\sigma_z^2)]$$

$$+\exp[-(z+H+2mL)^2/(2\sigma_z^2)]$$

$$+\exp[-(z+H-2mL)^2/(2\sigma_z^2)]$$

$$+\exp[-(z-H+2mL)^2/(2\sigma_z^2)]\}$$

C = concentration of emissions, in g/m³, at any receptor located:

x meters downwind from the emission source point

y meters crosswind from the emission plume centerline

z meters above ground level

Q = source pollutant emission rate, in g/s

u = horizontal wind velocity along the plume centerline, m/s

H = height of emission plume centerline above ground level, in m

σ_z = vertical standard deviation of the emission distribution, in m

σ_y = horizontal standard deviation of the emission distribution, in m

L = height from ground level to bottom of the inversion aloft, in m

exp = the exponential function

The above equation not only includes upward reflection from the ground, it also includes downward reflection from the bottom of any inversion lid present in the atmosphere.

The sum of the four exponential terms in g_3 converges to a final value quite rapidly. For most cases, the summation of the series with $m = 1$, $m = 2$ and $m = 3$ will provide an adequate solution.

σ_z and σ_y are functions of the atmospheric stability class (i.e., a measure of the turbulence in the ambient atmosphere) and of the downwind distance to the receptor. The two most important variables affecting the degree of pollutant emission dispersion obtained are the height of the emission source point and the degree of atmospheric turbulence. The more turbulence, the better the degree of dispersion.

The resulting calculations for air pollutant concentrations are often expressed as an air pollutant concentration contour map in order to show the spatial variation in contaminant levels over a wide area under study. In this way the contour lines can overlay sensitive receptor locations and reveal the spatial relationship of air pollutants to areas of interest.

Whereas older models rely on stability classes for the determination of σ_y and σ_z, more recent models increasingly rely on the Monin-Obukhov similarity theory to derive these parameters.

Briggs Plume Rise Equations

The Gaussian air pollutant dispersion equation (discussed above) requires the input of H which is the pollutant plume's centerline height above ground level—and H is the sum of H_s (the actual physical height of the pollutant plume's emission source point) plus ΔH (the plume rise due the plume's buoyancy).

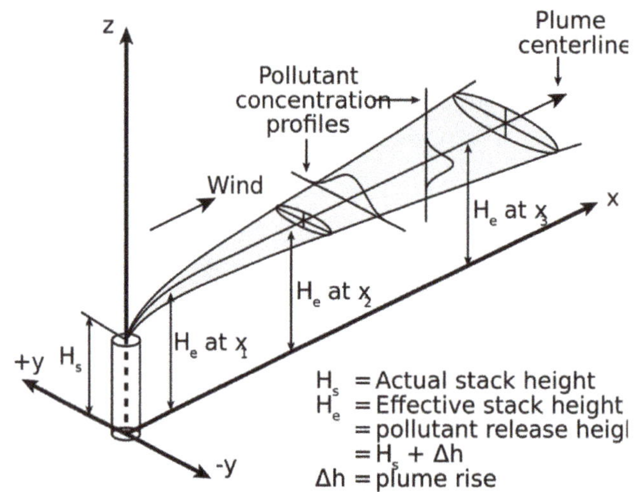

Visualization of a buoyant Gaussian air pollutant dispersion plume

To determine ΔH, many if not most of the air dispersion models developed between the late 1960s and the early 2000s used what are known as "the Briggs equations." G.A. Briggs first published his plume rise observations and comparisons in 1965. In 1968, at a symposium sponsored by CONCAWE (a Dutch organization), he compared many of the plume rise models then available in the literature. In that same year, Briggs also wrote the section of the publication edited by Slade dealing with the comparative analyses of plume rise models. That was followed in 1969 by his classical critical review of the entire plume rise literature, in which he proposed a set of plume rise equations which have become widely known as "the Briggs equations". Subsequently, Briggs modified his 1969 plume rise equations in 1971 and in 1972.

Briggs divided air pollution plumes into these four general categories:

- Cold jet plumes in calm ambient air conditions
- Cold jet plumes in windy ambient air conditions
- Hot, buoyant plumes in calm ambient air conditions
- Hot, buoyant plumes in windy ambient air conditions

Briggs considered the trajectory of cold jet plumes to be dominated by their initial velocity momentum, and the trajectory of hot, buoyant plumes to be dominated by their buoyant momentum to the extent that their initial velocity momentum was relatively unimportant. Although Briggs proposed plume rise equations for each of the above plume categories, *it is important to emphasize that "the Briggs equations" which become widely used are those that he proposed for bent-over, hot buoyant plumes.*

In general, Briggs's equations for bent-over, hot buoyant plumes are based on observations and data involving plumes from typical combustion sources such as the flue gas stacks from steam-generating boilers burning fossil fuels in large power plants. Therefore, the stack exit velocities were probably in the range of 20 to 100 ft/s (6 to 30 m/s) with exit temperatures ranging from 250 to 500 °F (120 to 260 °C).

A logic diagram for using the Briggs equations to obtain the plume rise trajectory of bent-over buoyant plumes is presented below:

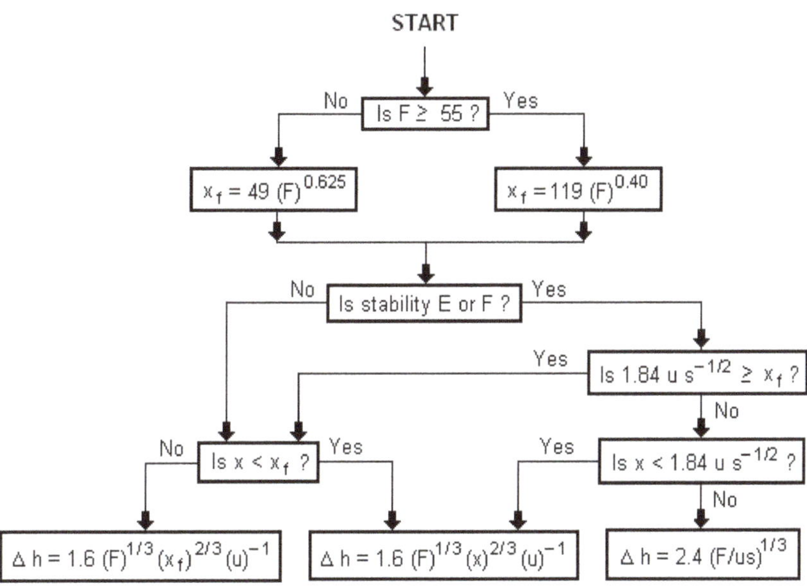

where:

Δh = plume rise, in m

F = buoyancy factor, in m^4s^{-3}

x = downwind distance from plume source, in m

$\mathbf{x_f}$ = downwind distance from plume source to point of maximum plume rise, in m

u = windspeed at actual stack height, in m/s

s = stability parameter, in s^{-2}

The above parameters used in the Briggs' equations are discussed in Beychok's book.

Outline of Air Pollution Dispersion

Air pollution dispersion – distribution of air pollution into the atmosphere. Air pollution is the introduction of particulates, biological molecules, or other harmful materials into Earth's atmosphere, causing disease, death to humans, damage to other living organisms such as food crops, or the natural or built environment. Air pollution may come from anthropogenic or natural sources. Dispersion refers to what happens to the pollution during and after its introduction; understanding this may help in identifying and controlling it. Air pollution dispersion has become the focus of environmental conservationists and governmental environmental protection agencies (local, state, province and national) of many countries (which have adopted and used much of the terminology of this field in their laws and regulations) regarding air pollution control.

Air Pollution Emission Plumes

- Buoyant plumes — Plumes which are lighter than air because they are at a higher temperature and lower density than the ambient air which surrounds them, or because they are at about the same temperature as the ambient air but have a lower molecular weight and hence lower density than the ambient air. For example, the emissions from the flue gas stacks of industrial furnaces are buoyant because they are considerably warmer and less dense than the ambient air. As another example, an emission plume of methane gas at ambient air temperatures is buoyant because methane has a lower molecular weight than the ambient air.
- Dense gas plumes — Plumes which are heavier than air because they have a higher density than the surrounding ambient air. A plume may have a higher density than air because it has a higher molecular weight than air (for example, a plume of carbon dioxide). A plume may also have a higher density than air if the plume is at a much lower temperature than the air. For example, a plume of evaporated gaseous methane from an accidental release of liquefied natural gas (LNG) may be as cold as -161 °C.
- Passive or neutral plumes — Plumes which are neither lighter or heavier than air.

Air Pollution Dispersion Models

There are five types of air pollution dispersion models, as well as some hybrids of the five types:

- Box model — The box model is the simplest of the model types. It assumes the airshed (i.e., a given volume of atmospheric air in a geographical region) is in the shape of a box. It also assumes that the air pollutants inside the box are homogeneously distributed and uses that assumption to estimate the average pollutant concentrations anywhere within the airshed. Although useful, this model is very limited in its ability to accurately predict dispersion of air pollutants over an airshed because the assumption of homogeneous pollutant distribution is much too simple.
- Gaussian model — The Gaussian model is perhaps the oldest (circa 1936) and perhaps the most commonly used model type. It assumes that the air pollutant dispersion has a Gaussian distribution, meaning that the pollutant distribution has a normal probability distribution. Gaussian models are most often used for predicting the dispersion of continuous, buoyant air pollution plumes originating from ground-level or elevated sources. Gaussian models may also be used for predicting the dispersion of non-continuous air pollution plumes (called *puff models*). The primary algorithm used in Gaussian modeling is the *Generalized Dispersion Equation For A Continuous Point-Source Plume.*

- Lagrangian model — a Lagrangian dispersion model mathematically follows pollution plume parcels (also called particles) as the parcels move in the atmosphere and they model the motion of the parcels as a random walk process. The Lagrangian model then calculates the air pollution dispersion by computing the statistics of the trajectories of a large number of the pollution plume parcels. A Lagrangian model uses a moving frame of reference as the parcels move from their initial location. It is said that an observer of a Lagrangian model follows along with the plume.
- Eulerian model — an Eulerian dispersions model is similar to a Lagrangian model in that it also tracks the movement of a large number of pollution plume parcels as they move from their initial location. The most important difference between the two models is that the Eulerian model uses a fixed three-dimensional Cartesian grid as a frame of reference rather than a moving frame of reference. It is said that an observer of an Eulerian model watches the plume go by.
- Dense gas model — Dense gas models are models that simulate the dispersion of dense gas pollution plumes (i.e., pollution plumes that are heavier than air). The three most commonly used dense gas models are:
  - The DEGADIS model developed by Dr. Jerry Havens and Dr. Tom Spicer at the University of Arkansas under commission by the US Coast Guard and US EPA.
  - The SLAB model developed by the Lawrence Livermore National Laboratory funded by the US Department of Energy, the US Air Force and the American Petroleum Institute.
  - The HEGADAS model developed by Shell Oil's research division.

Air Pollutant Emission

Air pollution emission source

- Types of air pollutant emission sources – named for their characteristics
  - Sources, by shape – there are four basic shapes which an emission source may have. They are:
    - ☐ Point source – single, identifiable source of air pollutant emissions (for example, the emissions from a combustion furnace flue gas stack). Point sources are also characterized as being either elevated or at ground-level. A point source has no geometric dimensions.
    - ☐ Line source – one-dimensional source of air pollutant emissions (for example, the emissions from the vehicular traffic on a roadway).
    - ☐ Area source – two-dimensional source of diffuse air pollutant emissions (for example, the emissions from a forest fire, a landfill or the evaporated vapors from a large spill of volatile liquid).
    - ☐ Volume source – three-dimensional source of diffuse air pollutant emissions. Essentially, it is an area source with a third (height) dimension (for example, the fugitive gaseous emissions from piping flanges, valves and other equipment at various heights within industrial facilities such as oil refineries and petrochemical plants). Another example would be the emissions from an automobile paint shop with multiple roof vents or multiple open windows.
  - Sources, by motion
    - ☐ Stationary source – flue gas stacks are examples of stationary sources
    - ☐ Mobile source – buses are examples of mobile sources
  - Sources, by urbanization level – whether the source is within a city or not is relevant in that urban areas constitute a so-called *heat island* and the heat rising from an urban area causes the atmosphere above an urban area to be more turbulent than the atmosphere above a rural area
    - ☐ Urban source – emission is in an urban area
    - ☐ Rural source – emission is in a rural area
  - Sources, by elevation
    - ☐ Surface or ground-level source
    - ☐ Near surface source
    - ☐ Elevated source
  - Sources, by duration
    - ☐ Puff or intermittent source – short term sources (for example, many acci-

dental emission releases are short term puffs)

- ☐ Continuous source – long term source (for example, most flue gas stack emissions are continuous)

Characterization of Atmospheric Turbulence

Effect of turbulence on dispersion – turbulence increases the entrainment and mixing of unpolluted air into the plume and thereby acts to reduce the concentration of pollutants in the plume (i.e., enhances the plume dispersion). It is therefore important to categorize the amount of atmospheric turbulence present at any given time...

The Pasquill Atmospheric Stability Classes

Pasquill atmospheric stability classes – oldest and, for a great many years, the most commonly used method of categorizing the amount of atmospheric turbulence present was the method developed by Pasquill in 1961. He categorized the atmospheric turbulence into six stability classes named A, B, C, D, E and F with class A being the most unstable or most turbulent class, and class F the most stable or least turbulent class. Table 1 lists the six classes and Table 2 provides the meteorological conditions that define each class.

Table 1: The Pasquill stability classes

Stability class	**Definition**		**Stability class**	**Definition**
A	very unstable		D	neutral
B	unstable		E	slightly stable
C	slightly unstable		F	stable

Table 2: Meteorological conditions that define the Pasquill stability classes

Surface windspeed		**Daytime incoming solar radiation**			**Nighttime cloud cover**	
m/s	mi/h	Strong	Moderate	Slight	> 50%	< 50%
< 2	< 5	A	A – B	B	E	F
2 – 3	5 – 7	A – B	B	C	E	F
3 – 5	7 – 11	B	B – C	C	D	E
5 – 6	11 – 13	C	C – D	D	D	D
> 6	> 13	C	D	D	D	D
Note: Class D applies to heavily overcast skies, at any windspeed day or night						

Data Availability

- Historical stability class data – known as the Stability Array (STAR) data, for sites within the USA can be purchased from the National Climatic Data Center (NCDC).

Advanced Methods of Categorizing Atmospheric Turbulence

Advanced air pollution dispersion models – they do not categorize atmospheric turbulence by using the simple meteorological parameters commonly used in defining the six Pasquill classes as shown in Table 2 above. The more advanced models use some form of Monin-Obukhov similarity theory. Some examples include:

- AERMOD – US EPA's most advanced model, no longer uses the Pasquill stability classes to categorize atmospheric turbulence. Instead, it uses the surface roughness length and the Monin-Obukhov length.
- ADMS 4, – United Kingdom's most advanced model, uses the Monin-Obukhov length, the boundary layer height and the windspeed to categorize the atmospheric turbulence.

Miscellaneous Other Terminology

- Building effects or downwash: When an air pollution plume flows over nearby buildings or other structures, turbulent eddies are formed in the downwind side of the building. Those eddies cause a plume from a stack source located within about five times the height of a nearby building or structure to be forced down to the ground much sooner than it would if a building or structure were not present. The effect can greatly increase the resulting nearby ground-level pollutant concentrations downstream of the building or structure. If the pollutants in the plume are subject to depletion by contact with the ground (particulates, for example), the concentration increase just downstream of the building or structure will decrease the concentrations further downstream.
- Deposition of the pollution plume components to the underlying surface can be defined as either dry or wet deposition:
    - Dry deposition is the removal of gaseous or particulate material from the pollution plume by contact with the ground surface or vegetation (or even water surfaces) through transfer processes such as absorption and gravitational sedimentation. This may be calculated by means of a *deposition velocity*, which is related to the resistance of the underlying surface to the transfer.
    - Wet deposition is the removal of pollution plume components by the action of rain. The wet deposition of radionuclides in a pollution plume by a burst of rain often forms so called *hot spots* of radioactivity on the underlying surface.
- Inversion layers: Normally, the air near the Earth's surface is warmer than the air above it because the atmosphere is heated from below as solar radiation warms the Earth's surface, which in turn then warms the layer of the atmosphere directly above it. Thus, the atmospheric temperature normally decreases with increasing altitude. However, under certain meteorological conditions, atmospheric layers may form in which the temperature increases with increasing altitude. Such layers are called inversion layers. When such a layer forms at the Earth's surface, it is called a surface inversion. When an inversion layer forms at some distance above the earth, it is called an inversion aloft (sometimes referred to as a *capping inversion*). The air within an inversion aloft is very stable with very little vertical motion. Any rising parcel of air within the inversion soon expands, thereby adiabatically cooling to a lower temperature than the surrounding air and the parcel stops rising. Any sinking parcel soon compresses adiabatically to a higher temperature than the surrounding

air and the parcel stops sinking. Thus, any air pollution plume that enters an inversion aloft will undergo very little vertical mixing unless it has sufficient momentum to completely pass through the inversion aloft. That is one reason why an inversion aloft is sometimes called a capping inversion.

- Mixing height: When an inversion aloft is formed, the atmospheric layer between the Earth's surface and the bottom of the inversion aloft is known as the mixing layer and the distance between the Earth's surface and the bottom of inversion aloft is known as the mixing height. Any air pollution plume dispersing beneath an inversion aloft will be limited in vertical mixing to that which occurs beneath the bottom of the inversion aloft (sometimes called the *lid*). Even if the pollution plume penetrates the inversion, it will not undergo any further significant vertical mixing. As for a pollution plume passing completely through an inversion layer aloft, that rarely occurs unless the pollution plume's source stack is very tall and the inversion lid is fairly low.

Roadway Air Dispersion Modeling

Roadway air dispersion is applied to highway segments

Roadway air dispersion modeling is the study of air pollutant transport from a roadway or other linear emitter. Computer models are required to conduct this analysis, because of the complex variables involved, including vehicle emissions, vehicle speed, meteorology, and terrain geometry. Line source dispersion has been studied since at least the 1960s, when the regulatory framework in the United States began requiring quantitative analysis of the air pollution consequences of major roadway and airport projects. By the early 1970s this subset of atmospheric dispersion models were being applied to real world cases of highway planning, even including some controversial court cases.

How the Model Works

The basic concept of the roadway air dispersion model is to calculate air pollutant levels in the vicinity of a highway or arterial roadway by considering them as line sources. The model takes into account source characteristics such as traffic volume, vehicle speeds, truck mix, and fleet emission controls; in addition, the roadway geometry, surrounding terrain and local meteorology are addressed. For example, many air quality standards require that certain near worst case meteorological conditions be applied.

The calculations are sufficiently complex that a computer model is essential to arrive at authoritative results, although workbook type manuals have been developed as screening techniques. In some cases where results must be refereed (such as legal cases), model validation may be needed with field test data in the local setting; this step is not usually warranted, because the best models have been extensively validated over a wide spectrum of input data variables.

The product of the calculations is usually a set of isopleths or mapped contour lines either in plan view or cross sectional view. Typically these might be stated as concentrations of carbon monoxide, total reactive hydrocarbons, oxides of nitrogen, particulate or benzene. The air quality scientist can run the model successively to study techniques of reducing adverse air pollutant concentrations (for example, by redesigning roadway geometry, altering speed controls or limiting certain types of trucks). The model is frequently utilized in an Environmental Impact Statement involving a major new roadway or land use change which will induce new vehicular traffic.

History

The logical building block for this theory was the use of the Gaussian air pollutant dispersion equation for point sources. One of the early point source air pollutant plume dispersion equations was derived by Bosanquet and Pearson in 1936. Their equation did not include the effect of ground reflection of the pollutant plume. Sir Graham Sutton derived a point source air pollutant plume dispersion equation in 1947 which included the assumption of Gaussian distribution for the vertical and crosswind dispersion of the plume and also addressed the effect of ground reflection of the plume. Further advances were made by G. A. Briggs in model refinement and validation and by D.B. Turner for his user-friendly workbook that included screening calculations which do not require a computer.

In seeing the need to develop a line source model to approach the study of roadway air pollution, Michael Hogan and Richard Venti developed a closed form solution to integrating the point source equation in a series of publications.

The source of virtually all roadway air pollution emissions is the exhaust

While the ESL mathematical model was completed for a line source by 1970, model refinement resulted in a "strip source", emulating the horizontal extent of the roadway surface. This theory would be the precursor of area source dispersion models. But their focus was roadway simulation, so they proceeded with the development of a computer model by adding to the team Leda Patmore, a computer programmer in the field of atmospheric physics and satellite trajectory calculations. A working computer model was produced by late 1970; then the model was calibrated with carbon monoxide field measurements targeting from traffic on U.S. Route 101 in Sunnyvale, California.

The ESL model received endorsement from the U.S. Environmental Protection Agency (EPA) in the form of a major grant to validate the model using actual roadway tests of tracer gas sulfur hexafluoride dispersion. That gas was chosen since it does not occur naturally or in vehicular emissions and provides a unique tracer for such dispersion studies. Part of the Environmental Protection Agency's motives may have been to bring the model into public domain. After a successful validation through the EPA research, the model was soon put to use in a variety of settings to forecast air pollution levels in the vicinity of roadways. The ESL group applied the model to the U.S. Route 101 bypass project in Cloverdale, California, the extension of Interstate 66 through Arlington, Virginia, the widening of the New Jersey Turnpike through Raritan and East Brunswick, New Jersey, and several transportation projects in Boston for the Boston Transportation Planning Review.

By the early 1970s at least two other research groups were known to be actively developing some type to roadway air dispersion model: the Environmental Research and Technology group of Lexington, Massachusetts and Caltrans headquarters in Sacramento, California. The Caline model of Caltrans borrowed some of the technology from the ESL Inc. group, since Caltrans funded some of the early model application work in Cloverdale and other locations and was given rights to use parts of their model.

The Theory

The resulting solution for an infinite line source is:

$$\chi = \int_0^{\infty} \frac{q}{\pi\left(ucdx^2\right)\left(cos\alpha\right)}\left(\exp\frac{y^2}{2c^2x^2}\right)dx$$

where:

x is the distance from the observer to the roadway

y is the height of the observer

u is the mean wind speed

α is the angle of tilt of the line source relative to the reference frame

c and d are the standard deviation of horizontal and vertical wind directions (measured in radians) respectively.

This equation was integrated into a closed form solution using the error function (erf), and variations in geometry can be performed to include the full infinite line, line segment, elevated line, or arc made from segments. In any case one can calculate three-dimensional contours of resulting air pollutant concentrations and use the mathematical model to study alternative roadway designs, various assumptions of worst case meteorology or varying traffic conditions (for example, variations in truck mix, fleet emission controls, or vehicle speed).

The ESL research group also extended their model by introducing the area source concept of a vertical strip to simulate the mixing zone on the highway produced by vehicle turbulence. This model too was validated in 1971 and showed good correlation with field test data.

Example Applications of the Model

Roadway air dispersion modeling is also done for curved roadways-North-South Express Highway, Malaysia

There were several early applications of the model in somewhat dramatic cases. In 1971 the Arlington Coalition on Transportation (ACT) was the plaintiff in an action against the Virginia Highway Commission over the extension of Interstate 66 through Arlington, Virginia, having filed a suit in the federal district court. The ESL model was used to produce calculations of air quality in the vicinity of the proposed highway. ACT won this case after a decision by the U.S. Fourth Circuit Court of Appeals. The court paid special attention to the plaintiff's expert calculations and testimony projecting that air quality levels would violate Federal ambient air quality standards as set forth in the Clean Air Act.

A second contentious case took place in East Brunswick, New Jersey where the New Jersey Turnpike Authority planned a major widening of the Turnpike. Again the roadway air dispersion model was employed to predict levels of air pollution for residences, schools and parks near the Turnpike. After an initial hearing in Superior Court where the ESL model results were set forth, the judge ordered the Turnpike Authority to negotiate with the plaintiff, Concerned Citizens of East Brunswick and develop air quality mitigation for the adverse effects. The Turnpike Authority hired ERT as its expert, and the two research teams negotiated a settlement to this case using the newly created roadway air dispersion models.

More Recent Model Refinements

The CALINE3 model is a steady-state Gaussian dispersion model designed to determine air pollution concentrations at receptor locations downwind of highways located in relatively uncomplicated terrain. CALINE3 is incorporated into the more elaborate CAL3QHC and CAL3QHCR models. CALINE3 is in widespread use due to its user friendly nature and promotion in governmental circles, but it falls short of analyzing the complexity of cases addressed by the original Hogan-Venti model. CAL3QHC and CAL3QHCR models are available in the Fortran programming language. They have options to model either particulate matter or carbon monoxide, and include algorithms to simulate queued traffic at signalized intersections .

In addition, several more recent models have been developed that employ non-steady state Lagrangian puff algorithms. The HYROAD dispersion model has been developed through the National Cooperative Highway Research Program's Project 25-06, incorporating ROADWAY-2 model puff and steady-state plume algorithms (Rao et al., 2002).

The TRAQSIM model, developed as part of a Ph.D dissertation with support by the U.S. Department of Transportation's Volpe National Transportation Systems Center's Air Quality Facility is currently under the care of Wyle. The model incorporates dynamic vehicle behavior with a non-steady state Gaussian puff algorithm. Unlike HYROAD, TRAQSIM combines traffic simulation, second-by-second modal emissions, and Gaussian puff dispersion into a fully integrated system (a true simulation) that models individual vehicles as discrete moving sources. TRAQSIM was developed as a next generation model to be the successor to the current CALINE3 and CAL3QHC regulatory models. The next step in the development of TRAQSIM is to incorporate methods to model the dispersion of particulate matter (PM) and hazardous air pollutants (HAPs).

Several models have been developed that handle complex urban meteorology resulting from urban canyons and highway configurations. The earliest such model development (1968-1970) was by the Air Pollution Control Office of the U.S. EPA in conjunction with New York City. The model was successfully applied to the Spadina Expressway in Toronto by Jack Fensterstock of the New York City Department of Air Resources,. Other examples include the Turner-Fairbank Highway Research Center's Canyon Plume Box model, now in version 3 (CPB-3), the National Environmental Research Institute of Denmark's Operational Street Pollution Model (OSPM), and the MICRO-CALGRID model, which includes photochemistry, allowing for both primary and secondary species to be modeled. Cornell University's CTAG model, which resolves vehicle-induced turbulence (VIT), road-induced turbulence (RIT), chemical transformation and aerosol dynamics of air pollutants using turbulence reacting flow models. The CTAG model has also been applied to characterize highway-building environments and study effects of vegetation barriers on near-road air pollution.

Recent Applications in Legal Cases

Recent health literature indicating that residents near major roads face elevated rates of several adverse health outcomes has prompted legal dispute over the responsibility of transportation agencies to use roadway air dispersion models to characterize the impacts of new and expanded roadways, bus terminals, truck stops, and other sources.

Recently, the Sierra Club of Nevada sued the Nevada Department of Transportation and the Federal Highway Administration over its failure to assess the impact of the expansion of U.S. Route 95 in Las Vegas on neighborhood air quality. The Sierra Club asserted that a supplemental Environmental Impact Statement should be issued to address emissions of hazardous air pollutants and particulate matter from new motor vehicle traffic. The plaintiffs asserted that modeling tools were available, including the Environmental Protection Agency's MOBILE6.2 model, the CALINE3 dispersion model, and other relevant models. The defendants won in the U.S. District Court under Judge Philip Pro, who ruled that the transportation agencies had acted in a manner that was not "arbitrary and capricious," despite the agencies' technical arguments regarding the lack of available modeling tools being contradicted by a number of peer-reviewed studies published in scientific journals (e.g. Korenstein and Piazza, Journal of Environmental Health, 2002). On appeal to the U.S. Ninth Circuit, the Appeals Court stayed new construction on the highway pending the court's final decision. The Sierra Club and the defendants settled out of court, setting up a research program on the air quality impacts of U.S. Route 95 on nearby schools.

A number of other high-profile cases have prompted environmental groups to call for dispersion modeling to be used to assess the air quality impacts of new transportation projects on nearby communities, but to date state transportation agencies and the Federal Highways Administration has claimed that no tools are available, despite models and guidance available through EPA's Support Center for Regulatory Air Models (SCRAM).

Among the more contentious of cases the Detroit Intermodal Freight Terminal and Detroit River International Crossing (Michigan, USA), and the expansion of Interstate 70 East in Denver (Colorado, USA).

In all of these cases, community-based organizations have asserted that modeling tools are available, but transportation planning agencies have asserted that too much uncertainty exists in all of the steps. A major concern for community-based organizations has been transportation agencies' unwillingness to define the level of uncertainty that they are willing to tolerate in air quality analyses, how that compares to the Environmental Protection Agency's guideline on air quality models, which addresses uncertainty and accuracy in model use.

References

- Beychok, Milton R. (2005). Fundamentals Of Stack Gas Dispersion (4th ed.). author-published. ISBN 0-9644588-0-2.
- Turner, D.B. (1994). Workbook of atmospheric dispersion estimates: an introduction to dispersion modeling (2nd ed.). CRC Press. ISBN 1-56670-023-X.
- Beychok, Milton R. (2005). Fundamentals Of Stack Gas Dispersion (4th ed.). author-published. ISBN 0-9644588-0-2.
- Turner, D.B. (1994). Workbook of atmospheric dispersion estimates: an introduction to dispersion modeling (2nd ed.). CRC Press. ISBN 1-56670-023-X.

7

Air Quality, Measurement and Health Index

This chapter explains the measurements taken by the governments for air pollution. A continuous increase in Air Quality Index (AQI) ensures a large percentage of the population to experience severe heath effects. Every country has its own air quality index and some of these are the Air Quality Health Index (Canada), the Air Pollution Index (Malaysia), and the Pollutant Standards Index (Singapore). To improve the quality of air, we certainly need to implement the measurements taken by our governments.

Air Quality Index

An air quality index (AQI) is a number used by government agencies to communicate to the public how polluted the air currently is or how polluted it is forecast to become. As the AQI increases, an increasingly large percentage of the population is likely to experience increasingly severe adverse health effects. Different countries have their own air quality indices, corresponding to different national air quality standards. Some of these are the Air Quality Health Index (Canada), the Air Pollution Index (Malaysia), and the Pollutant Standards Index (Singapore).

Wildfires give ris-*****e to an elevated AQI in parts of Greece

Definition and Usage

Computation of the AQI requires an air pollutant concentration over a specified averaging period, obtained from an air monitor or model. Taken together, concentration and time represent the dose of the air pollutant. Health effects corresponding to a given dose are established by epidemiological research. Air pollutants vary in potency, and the function used to convert from air pollutant concentration to AQI varies by pollutant. Air quality index values are typically grouped into ranges. Each range is assigned a descriptor, a color code, and a standardized public health advisory.

An air quality measurement station in Edinburgh, Scotland

The AQI can increase due to an increase of air emissions (for example, during rush hour traffic or when there is an upwind forest fire) or from a lack of dilution of air pollutants. Stagnant air, often caused by an anticyclone, temperature inversion, or low wind speeds lets air pollution remain in a local area, leading to high concentrations of pollutants, chemical reactions between air contaminants and hazy conditions.

Signboard in Gulfton, Houston indicating an ozone watch

On a day when the AQI is predicted to be elevated due to fine particle pollution, an agency or public health organization might:

- advise sensitive groups, such as the elderly, children, and those with respiratory or cardiovascular problems to avoid outdoor exertion.
- declare an "action day" to encourage voluntary measures to reduce air emissions, such as using public transportation.
- recommend the use of masks to keep fine particles from entering the lungs

Woman wearing an air pollution mask in Beijing, China

During a period of very poor air quality, such as an air pollution episode, when the AQI indicates that acute exposure may cause significant harm to the public health, agencies may invoke emergency plans that allow them to order major emitters (such as coal burning industries) to curtail emissions until the hazardous conditions abate.

Most air contaminants do not have an associated AQI. Many countries monitor ground-level ozone, particulates, sulfur dioxide, carbon monoxide and nitrogen dioxide, and calculate air quality indices for these pollutants.

The definition of the AQI in a particular nation reflects the discourse surrounding the development of national air quality standards in that nation. A website allowing government agencies anywhere in the world to submit their real-time air monitoring data for display using a common definition of the air quality index has recently become available.

Indices by Location

Canada

Air quality in Canada has been reported for many years with provincial Air Quality Indices (AQIs). Significantly, AQI values reflect air quality management objectives, which are based on the lowest achievable emissions rate, and not exclusively concern for human health. The Air Quality Health Index or (AQHI) is a scale designed to help understand the impact of air quality on health. It is a health protection tool used to make decisions to reduce short-term exposure to air pollution by adjusting activity levels during increased levels of air pollution. The Air Quality Health Index also provides advice on how to improve air quality by proposing behavioural change to reduce the environmental footprint. This index pays particular attention to people who are sensitive to air pollution. It provides them with advice on how to protect their health during air quality levels associated with low, moderate, high and very high health risks.

The Air Quality Health Index provides a number from 1 to 10+ to indicate the level of health risk associated with local air quality. On occasion, when the amount of air pollution is abnormally high, the number may exceed 10. The AQHI provides a local air quality current value as well as a local air quality maximums forecast for today, tonight, and tomorrow, and provides associated health advice.

Risk: Low (1–3) Moderate (4–6) High (7–10) Very high (above 10)

Health Risk	Air Quality Health Index	Health Messages	
		At Risk population	***General Population**
Low	**1–3**	**Enjoy** your usual outdoor activities.	**Ideal** air quality for outdoor activities
Moderate	**4–6**	**Consider reducing** or rescheduling strenuous activities outdoors if you are experiencing symptoms.	**No need to modify** your usual outdoor activities unless you experience symptoms such as coughing and throat irritation.
High	**7–10**	**Reduce** or reschedule strenuous activities outdoors. Children and the elderly should also take it easy.	**Consider reducing** or rescheduling strenuous activities outdoors if you experience symptoms such as coughing and throat irritation.
Very high	**Above 10**	**Avoid** strenuous activities outdoors. Children and the elderly should also avoid outdoor physical exertion.	**Reduce** or reschedule strenuous activities outdoors, especially if you experience symptoms such as coughing and throat irritation.

Hong Kong

On the 30th December 2013 Hong Kong replaced the Air Pollution Index with a new index called the *Air Quality Health Index*. This index is on a scale of 1 to 10+ and considers four air pollutants: ozone; nitrogen dioxide; sulphur dioxide and particulate matter (including PM10 and PM2.5). For any given hour the AQHI is calculated from the sum of the percentage excess risk of daily hospital admissions attributable to the 3-hour moving average concentrations of these four pollutants. The AQHIs are grouped into five AQHI health risk categories with health advice provided:

Health risk category	AQHI
Low	1
	2
	3
Medium	4
	5
	6

High	7
Very High	8
	9
	10
Serious	10+

Each of the health risk categories has advice with it. At the *low* and *moderate* levels the public are advised that they can continue normal activities. For the *high* category, children, the elderly and people with heart or respiratory illnesses are advising to reduce outdoor physical exertion. Above this (*very high* or *serious*) the general public are also advised to reduce or avoid outdoor physical exertion.

Mainland China

China's Ministry of Environmental Protection (MEP) is responsible for measuring the level of air pollution in China. As of 1 January 2013, MEP monitors daily pollution level in 163 of its major cities. The API level is based on the level of 6 atmospheric pollutants, namely sulfur dioxide (SO_2), nitrogen dioxide (NO_2), suspended particulates smaller than 10 μm in aerodynamic diameter (PM_{10}), suspended particulates smaller than 2.5 μm in aerodynamic diameter ($PM_{2.5}$), carbon monoxide (CO), and ozone (O_3) measured at the monitoring stations throughout each city.

AQI Mechanics

An individual score (IAQI) is assigned to the level of each pollutant and the final AQI is the highest of those 6 scores. The pollutants can be measured quite differently. $PM_{2.5}$、PM_{10} concentration are measured as average per 24h. SO_2, NO_2, O_3, CO are measured as average per hour. The final API value is calculated per hour according to a formula published by the MEP.

The scale for each pollutant is non-linear, as is the final AQI score. Thus an AQI of 100 does not mean twice the pollution of AQI at 50, nor does it mean twice as harmful. While an AQI of 50 from day 1 to 182 and AQI of 100 from day 183 to 365 does provide an annual average of 75, it does *not* mean the pollution is acceptable even if the benchmark of 100 is deemed safe. This is because the benchmark is a 24-hour target. The annual average must match against the annual target. It is entirely possible to have safe air every day of the year but still fail the annual pollution benchmark.

AQI and Health Implications (HJ 663-2012)

AQI	Air Pollution Level	Health Implications
0–50	Excellent	No health implications.
51–100	Good	Few hypersensitive individuals should reduce outdoor exercise.
101–150	Lightly Polluted	Slight irritations may occur, individuals with breathing or heart problems should reduce outdoor exercise.
151–200	Moderately Polluted	Slight irritations may occur, individuals with breathing or heart problems should reduce outdoor exercise.

201–300	Heavily Polluted	Healthy people will be noticeably affected. People with breathing or heart problems will experience reduced endurance in activities. These individuals and elders should remain indoors and restrict activities.
300+	Severely Polluted	Healthy people will experience reduced endurance in activities. There may be strong irritations and symptoms and may trigger other illnesses. Elders and the sick should remain indoors and avoid exercise. Healthy individuals should avoid outdoor activities.

India

The Minister for Environment, Forests & Climate Change Shri Prakash Javadekar launched The National Air Quality Index (AQI) in New Delhi on 17 September 2014 under the Swachh Bharat Abhiyan. It is outlined as 'One Number- One Colour-One Description' for the common man to judge the air quality within his vicinity. The index constitutes part of the Government's mission to introduce the culture of cleanliness. Institutional and infrastructural measures are being undertaken in order to ensure that the mandate of cleanliness is fulfilled across the country and the Ministry of Environment, Forests & Climate Change proposed to discuss the issues concerned regarding quality of air with the Ministry of Human Resource Development in order to include this issue as part of the sensitisation programme in the course curriculum.

While the earlier measuring index was limited to three indicators, the current measurement index had been made quite comprehensive by the addition of five additional parameters. Under the current measurement of air quality there are 8 parameters . The initiatives undertaken by the Ministry recently aimed at balancing environment and conservation and development as air pollution has been a matter of environmental and health concerns, particularly in urban areas.

The Central Pollution Control Board along with State Pollution Control Boards has been operating National Air Monitoring Program (NAMP) covering 240 cities of the country. In addition, continuous monitoring systems that provide data on near real-time basis are also installed in a few cities. They provide information on air quality in public domain in simple linguistic terms that is easily understood by a common person. Air Quality Index (AQI) is one such tool for effective dissemination of air quality information to people. As such an Expert Group comprising medical professionals, air quality experts, academia, advocacy groups, and SPCBs was constituted and a technical study was awarded to IIT Kanpur. IIT Kanpur and the Expert Group recommended an AQI scheme in 2014.

There are six AQI categories, namely Good, Satisfactory, Moderately polluted, Poor, Very Poor, and Severe. The proposed AQI will consider eight pollutants (PM_{10}, $PM_{2.5}$, NO_2, SO_2, CO, O_3, NH_3, and Pb) for which short-term (up to 24-hourly averaging period) National Ambient Air Quality Standards are prescribed. Based on the measured ambient concentrations, corresponding standards and likely health impact, a sub-index is calculated for each of these pollutants. The worst sub-index reflects overall AQI. Associated likely health impacts for different AQI categories and pollutants have been also been suggested, with primary inputs from the medical expert members of the group. The AQI values and corresponding ambient concentrations (health breakpoints) as well as associated likely health impacts for the identified eight pollutants are as follows:

AQI Category, Pollutants and Health Breakpoints								
AQI Category (Range)	**PM_{10} (24hr)**	**$PM_{2.5}$ (24hr)**	**NO_2 (24hr)**	**O_3 (8hr)**	**CO (8hr)**	**SO_2 (24hr)**	**NH_3 (24hr)**	**Pb (24hr)**
Good (0-50)	0-50	0-30	0-40	0-50	0-1.0	0-40	0-200	0-0.5
Satisfactory (51-100)	51-100	31-60	41-80	51-100	1.1-2.0	41-80	201-400	0.5-1.0
Moderately polluted (101-200)	101-250	61-90	81-180	101-168	2.1-10	81-380	401-800	1.1-2.0
Poor (201-300)	251-350	91-120	181-280	169-208	10-17	381-800	801-1200	2.1-3.0
Very poor (301-400)	351-430	121-250	281-400	209-748	17-34	801-1600	1200-1800	3.1-3.5
Severe (401-500)	430+	250+	400+	748+	34+	1600+	1800+	3.5+

AQI	Associated Health Impacts
Good (0-50)	Minimal impact
Satisfactory (51-100)	May cause minor breathing discomfort to sensitive people.
Moderately polluted (101–200)	May cause breathing discomfort to people with lung disease such as asthma, and discomfort to people with heart disease, children and older adults.
Poor (201-300)	May cause breathing discomfort to people on prolonged exposure, and discomfort to people with heart disease.
Very poor (301-400)	May cause respiratory illness to the people on prolonged exposure. Effect may be more pronounced in people with lung and heart diseases.
Severe (401-500)	May cause respiratory impact even on healthy people, and serious health impacts on people with lung/heart disease. The health impacts may be experienced even during light physical activity.

Mexico

The air quality in Mexico City is reported in IMECAs. The IMECA is calculated using the measurements of average times of the chemicals ozone (O_3), sulphur dioxide (SO_2), nitrogen dioxide (NO_2), carbon monoxide (CO), particles smaller than 2.5 micrometers ($PM_{2.5}$), and particles smaller than 10 micrometers (PM_{10}).

Singapore

Singapore uses the Pollutant Standards Index to report on its air quality, with details of the calculation similar but not identical to that used in Malaysia and Hong Kong The PSI chart below is grouped by index values and descriptors, according to the National Environment Agency.

PSI	Descriptor	General Health Effects
0–50		None
51–100	Moderate	Few or none for the general population
101–200	Unhealthy	Mild aggravation of symptoms among susceptible persons i.e. those with underlying conditions such as chronic heart or lung ailments; transient symptoms of irritation e.g. eye irritation, sneezing or coughing in some of the healthy population.

201–300	Very Unhealthy	Moderate aggravation of symptoms and decreased tolerance in persons with heart or lung disease; more widespread symptoms of transient irritation in the healthy population.
301–400	Hazardous	Early onset of certain diseases in addition to significant aggravation of symptoms in susceptible persons; and decreased exercise tolerance in healthy persons.
Above 400	Hazardous	PSI levels above 400 may be life-threatening to ill and elderly persons. Healthy people may experience adverse symptoms that affect normal activity.

South Korea

The Ministry of Environment of South Korea uses the Comprehensive Air-quality Index (CAI) to describe the ambient air quality based on the health risks of air pollution. The index aims to help the public easily understand the air quality and protect people's health. The CAI is on a scale from 0 to 500, which is divided into six categories. The higher the CAI value, the greater the level of air pollution. Of values of the five air pollutants, the highest is the CAI value. The index also has associated health effects and a colour representation of the categories as shown below.

CAI	Description	Health Implications
0–50	Good	A level that will not impact patients suffering from diseases related to air pollution.
51–100	Moderate	A level that may have a meager impact on patients in case of chronic exposure.
101–150	Unhealthy for sensitive groups	A level that may have harmful impacts on patients and members of sensitive groups.
151–250	Unhealthy	A level that may have harmful impacts on patients and members of sensitive groups (children, aged or weak people), and also cause the general public unpleasant feelings.
251–500	Very unhealthy	A level that may have a serious impact on patients and members of sensitive groups in case of acute exposure.

The N Seoul Tower on Namsan Mountain in central Seoul, South Korea, is illuminated in blue, from sunset to 23:00 and 22:00 in winter, on days where the air quality in Seoul is 45 or less. During the spring of 2012, the Tower was lit up for 52 days, which is four days more than in 2011.

United Kingdom

The most commonly used air quality index in the UK is the *Daily Air Quality Index* recommended by the Committee on Medical Effects of Air Pollutants (COMEAP). This index has ten points, which are further grouped into 4 bands: low, moderate, high and very high. Each of the bands comes with advice for at-risk groups and the general population.

Air pollution banding	Value	Health messages for At-risk individuals	Health messages for General population
Low	1–3	Enjoy your usual outdoor activities.	Enjoy your usual outdoor activities.
Moderate	4–6	Adults and children with lung problems, and adults with heart problems, who experience symptoms, should consider reducing strenuous physical activity, particularly outdoors.	Enjoy your usual outdoor activities.
High	7–9	Adults and children with lung problems, and adults with heart problems, should reduce strenuous physical exertion, particularly outdoors, and particularly if they experience symptoms. People with asthma may find they need to use their reliever inhaler more often. Older people should also reduce physical exertion.	Anyone experiencing discomfort such as sore eyes, cough or sore throat should consider reducing activity, particularly outdoors.
Very High	10	Adults and children with lung problems, adults with heart problems, and older people, should avoid strenuous physical activity. People with asthma may find they need to use their reliever inhaler more often.	Reduce physical exertion, particularly outdoors, especially if you experience symptoms such as cough or sore throat.

The index is based on the concentrations of 5 pollutants. The index is calculated from the concentrations of the following pollutants: Ozone, Nitrogen Dioxide, Sulphur Dioxide, PM2.5 (particles with an aerodynamic diameter less than 2.5 μm) and PM10. The breakpoints between index values are defined for each pollutant separately and the overall index is defined as the maximum value of the index. Different averaging periods are used for different pollutants.

Index	Ozone, Running 8 hourly mean (μg/m³)	Nitrogen Dioxide, Hourly mean (μg/m³)	Sulphur Dioxide, 15 minute mean (μg/m³)	PM2.5 Particles, 24 hour mean (μg/m³)	PM10 Particles, 24 hour mean (μg/m³)
1	0-33	0-67	0-88	0-11	0-16
2	34-66	68-134	89-177	12-23	17-33
3	67-100	135-200	178-266	24-35	34-50
4	101-120	201-267	267-354	36-41	51-58
5	121-140	268-334	355-443	42-47	59-66
6	141-160	335-400	444-532	48-53	67-75
7	161-187	401-467	533-710	54-58	76-83
8	188-213	468-534	711-887	59-64	84-91
9	214-240	535-600	888-1064	65-70	92-100
10	≥ 241	≥ 601	≥ 1065	≥ 71	≥ 101

Europe

To present the air quality situation in European cities in a comparable and easily understandable way, all detailed measurements are transformed into a single relative figure: the Common Air Quality Index (or CAQI) Three different indices have been developed by Citeair to enable the comparison of three different time scale:.

- An hourly index, which describes the air quality today, based on hourly values and updated every hours,

- A daily index, which stands for the general air quality situation of yesterday, based on daily values and updated once a day,
- An annual index, which represents the city's general air quality conditions throughout the year and compare to European air quality norms. This index is based on the pollutants year average compare to annual limit values, and updated once a year.

However, the proposed indices and the supporting common web site www.airqualitynow.eu are designed to give a dynamic picture of the air quality situation in each city but not for compliance checking.

The Hourly and Daily Common Indices

These indices have 5 levels using a scale from 0 (very low) to > 100 (very high), it is a relative measure of the amount of air pollution. They are based on 3 pollutants of major concern in Europe: PM10, NO2, O3 and will be able to take into account to 3 additional pollutants (CO, PM2.5 and SO2) where data are also available.

The calculation of the index is based on a review of a number of existing air quality indices, and it reflects EU alert threshold levels or daily limit values as much as possible. In order to make cities more comparable, independent of the nature of their monitoring network two situations are defined:

- Background, representing the general situation of the given agglomeration (based on urban background monitoring sites),
- Roadside, being representative of city streets with a lot of traffic, (based on roadside monitoring stations)

The indices values are updated hourly (for those cities that supply hourly data) and yesterdays daily indices are presented.

Common Air Quality Index Legend:

Pollution	Index Value
Very low	0/25
Low	25/50
Medium	50/75
High	75/100
Very high	>100

The Common Annual Air Quality Index

The common annual air quality index provides a general overview of the air quality situation in a given city all the year through and regarding to the European norms.

It is also calculated both for background and traffic conditions but its principle of calculation is different from the hourly and daily indices. It is presented as a distance to a target index, this target being derived from the EU directives (annual air quality standards and objectives):

- If the index is higher than 1: for one or more pollutants the limit values are not met.
- If the index is below 1: on average the limit values are met.

The annual index is aimed at better taking into account long term exposure to air pollution based on distance to the target set by the EU annual norms, those norms being linked most of the time to recommendations and health protection set up by World Health Organisation.

United States

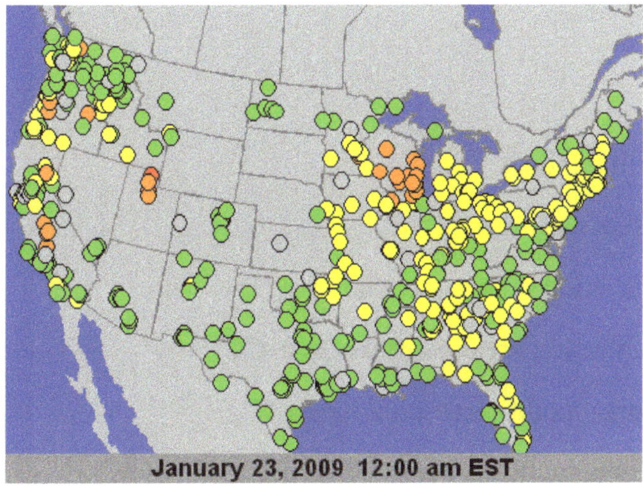

$PM_{2.5}$ 24-Hour AQI Loop, Courtesy US EPA

The United States Environmental Protection Agency (EPA) has developed an Air Quality Index that is used to report air quality. This AQI is divided into six categories indicating increasing levels of health concern. An AQI value over 300 represents hazardous air quality and below 50 the air quality is good.

Air Quality Index (AQI) Values	Levels of Health Concern	Colors
0 to 50	Good	Green
51 to 100	Moderate	Yellow
101 to 150	Unhealthy for Sensitive Groups	Orange
151 to 200	Unhealthy	Red
201 to 300	Very Unhealthy	Purple
301 to 500	Hazardous	Maroon

The AQI is based on the five "criteria" pollutants regulated under the Clean Air Act: ground-level ozone, particulate matter, carbon monoxide, sulfur dioxide, and nitrogen dioxide. The EPA has established National Ambient Air Quality Standards (NAAQS) for each of these pollutants in order to protect public health. An AQI value of 100 generally corresponds to the level of the NAAQS for the pollutant. The Clean Air Act (USA) (1990) requires EPA to review its National Ambient Air Quality Standards every five years to reflect evolving health effects information. The Air Quality Index is adjusted periodically to reflect these changes.

Computing the AQI

The air quality index is a piecewise linear function of the pollutant concentration. At the boundary between AQI categories, there is a discontinuous jump of one AQI unit. To convert from concentration to AQI this equation is used:

$$I = \frac{I_{high} - I_{low}}{C_{high} - C_{low}}(C - C_{low}) + I_{low}$$

where:

I = the (Air Quality) index,

C = the pollutant concentration,

C_{low} = the concentration breakpoint that is $\leq C$,

C_{high} = the concentration breakpoint that is $\geq C$,

I_{low} = the index breakpoint corresponding to C_{low},

I_{high} = the index breakpoint corresponding to C_{high}.

EPA's table of breakpoints is:

O_3 (ppb)	O_3 (ppb)	$PM_{2.5}$ (μg/m³)	PM_{10} (μg/m³)	CO (ppm)	SO_2 (ppb)	NO_2 (ppb)	AQI	AQI
C_{low} - C_{high} (avg)	C_{low} - C_{high} (avg)	C_{low} - C_{high} (avg)	C_{low} - C_{high} (avg)	C_{low} - C_{high} (avg)	C_{low} - C_{high} (avg)	C_{low} - C_{high} (avg)	I_{low} - I_{high}	**Category**
0-54 (8-hr)	-	0.0-12.0 (24-hr)	0-54 (24-hr)	0.0-4.4 (8-hr)	0-35 (1-hr)	0-53 (1-hr)	0-50	Good
55-70 (8-hr)	-	12.1-35.4 (24-hr)	55-154 (24-hr)	4.5-9.4 (8-hr)	36-75 (1-hr)	54-100 (1-hr)	51-100	Moderate
71-85 (8-hr)	125-164 (1-hr)	35.5-55.4 (24-hr)	155-254 (24-hr)	9.5-12.4 (8-hr)	76-185 (1-hr)	101-360 (1-hr)	101-150	Unhealthy for Sensitive Groups
86-105 (8-hr)	165-204 (1-hr)	55.5-150.4 (24-hr)	255-354 (24-hr)	12.5-15.4 (8-hr)	186-304 (1-hr)	361-649 (1-hr)	151-200	Unhealthy
106-200 (8-hr)	205-404 (1-hr)	150.5-250.4 (24-hr)	355-424 (24-hr)	15.5-30.4 (8-hr)	305-604 (24-hr)	650-1249 (1-hr)	201-300	Very Unhealthy
-	405-504 (1-hr)	250.5-350.4 (24-hr)	425-504 (24-hr)	30.5-40.4 (8-hr)	605-804 (24-hr)	1250-1649 (1-hr)	301-400	Hazardous
-	505-604 (1-hr)	350.5-500.4 (24-hr)	505-604 (24-hr)	40.5-50.4 (8-hr)	805-1004 (24-hr)	1650-2049 (1-hr)	401-500	

Suppose a monitor records a 24-hour average fine particle ($PM_{2.5}$) concentration of 12.0 micrograms per cubic meter. The equation above results in an AQI of:

$$\frac{50-0}{12.0-0}(12.0-0)+0=50,$$

corresponding to air quality in the "Good" range. To convert an air pollutant concentration to an AQI, EPA has developed a calculator.

If multiple pollutants are measured at a monitoring site, then the largest or "dominant" AQI value is reported for the location. The ozone AQI between 100 and 300 is computed by selecting the larger of the AQI calculated with a 1-hour ozone value and the AQI computed with the 8-hour ozone value.

8-hour ozone averages do not define AQI values greater than 300; AQI values of 301 or greater are calculated with 1-hour ozone concentrations. 1-hour SO_2 values do not define higher AQI values greater than 200. AQI values of 201 or greater are calculated with 24-hour SO_2 concentrations.

Real time monitoring data from continuous monitors are typically available as 1-hour averages. However, computation of the AQI for some pollutants requires averaging over multiple hours of data. (For example, calculation of the ozone AQI requires computation of an 8-hour average and computation of the $PM_{2.5}$ or PM_{10} AQI requires a 24-hour average.) To accurately reflect the current air quality, the multi-hour average used for the AQI computation should be centered on the current time, but as concentrations of future hours are unknown and are difficult to estimate accurately, EPA uses surrogate concentrations to estimate these multi-hour averages. For reporting the $PM_{2.5}$, PM_{10} and ozone air quality indices, this surrogate concentration is called the NowCast. The Nowcast is a particular type of weighted average that provides more weight to the most recent air quality data when air pollution levels are changing.

Public Availability of the AQI

Real time monitoring data and forecasts of air quality that are color-coded in terms of the air quality index are available from EPA's AirNow web site. Historical air monitoring data including AQI charts and maps are available at EPA's AirData website.

History of the AQI

The AQI made its debut in 1968, when the National Air Pollution Control Administration undertook an initiative to develop an air quality index and to apply the methodology to Metropolitan Statistical Areas. The impetus was to draw public attention to the issue of air pollution and indirectly push responsible local public officials to take action to control sources of pollution and enhance air quality within their jurisdictions.

Jack Fensterstock, the head of the National Inventory of Air Pollution Emissions and Control Branch, was tasked to lead the development of the methodology and to compile the air quality and emissions data necessary to test and calibrate resultant indices.

The initial iteration of the air quality index used standardized ambient pollutant concentrations to yield individual pollutant indices. These indices were then weighted and summed to form a single total air quality index. The overall methodology could use concentrations that are taken from ambient monitoring data or are predicted by means of a diffusion model. The concentrations were then converted into a standard statistical distribution with a preset mean and standard deviation. The resultant individual pollutant indices are assumed to be equally weighted, although values other than unity can be used. Likewise, the index can incorporate any number of pollutants although it was only used to combine SOx, CO, and TSP because of a lack of available data for other pollutants.

While the methodology was designed to be robust, the practical application for all metropolitan areas proved to be inconsistent due to the paucity of ambient air quality monitoring data, lack of agreement on weighting factors, and non-uniformity of air quality standards across geographical and political boundaries. Despite these issues, the publication of lists ranking metropolitan areas achieved the public policy objectives and led to the future development of improved indices and their routine application.

Acid rain

Acid rain is a rain or any other form of precipitation that is unusually acidic, meaning that it possesses elevated levels of hydrogen ions (low pH). It can have harmful effects on plants, aquatic animals and infrastructure. Acid rain is caused by emissions of sulfur dioxide and nitrogen oxide, which react with the water molecules in the atmosphere to produce acids. Some Governments have made efforts since the 1970s to reduce the release of sulfur dioxide and nitrogen oxide into the atmosphere with positive results. Nitrogen oxides can also be produced naturally by lightning strikes, and sulfur dioxide is produced by volcanic eruptions. The chemicals in acid rain can cause paint to peel, corrosion of steel structures such as bridges, and weathering of stone buildings and statues.

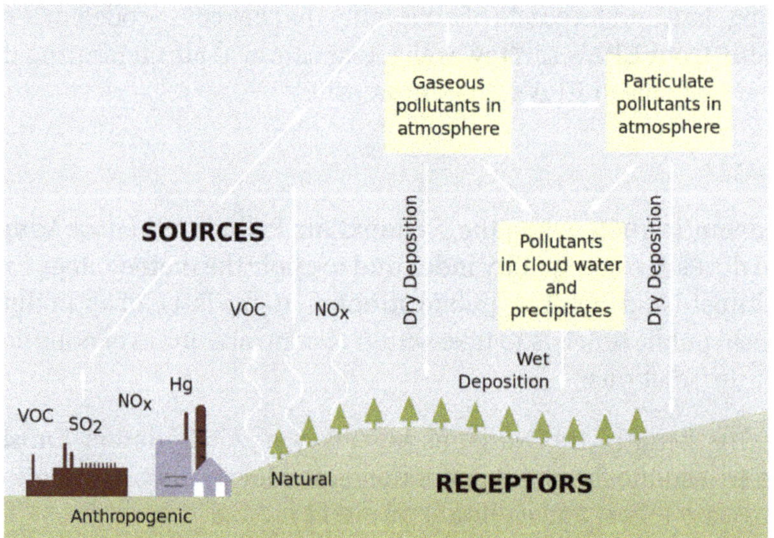

Processes involved in acid deposition (note that only SO_2 and NO_x play a significant role in acid rain).

Acid clouds can grow on SO_2 emissions from refineries, as seen here in Curaçao.

Definition

"Acid rain" is a popular term referring to the deposition of a mixture from wet (rain, snow, sleet, fog, cloudwater, and dew) and dry (acidifying particles and gases) acidic components. Distilled water, once carbon dioxide is removed, has a neutral pH of 7. Liquids with a pH less than 7 are acidic, and those with a pH greater than 7 are alkaline. "Clean" or unpolluted rain has an acidic pH, but usually no lower than 5.7, because carbon dioxide and water in the air react together to form carbonic acid, a weak acid according to the following reaction:

$$H_2O\ (l) + CO_2\ (g) \rightleftharpoons H_2CO_3\ (aq)$$

Carbonic acid then can ionize in water forming low concentrations of hydronium and carbonate ions:

$$H_2O\ (l) + H_2CO_3\ (aq) \rightleftharpoons HCO_3^-\ (aq) + H_3O^+\ (aq)$$

However, unpolluted rain can also contain other chemicals which affect its pH (acidity level). A common example is nitric acid produced by electric discharge in the atmosphere such as lightning. Acid deposition as an environmental issue would include additional acids to H_2CO_3.

History

The corrosive effect of polluted, acidic city air on limestone and marble was noted in the 17th century by John Evelyn, who remarked upon the poor condition of the Arundel marbles. Since the Industrial Revolution, emissions of sulfur dioxide and nitrogen oxides into the atmosphere have increased. In 1852, Robert Angus Smith was the first to show the relationship between acid rain and atmospheric pollution in Manchester, England.

Though acidic rain was discovered in 1853, it was not until the late 1960s that scientists began widely observing and studying the phenomenon. The term "acid rain" was coined in 1872 by Robert Angus Smith. Canadian Harold Harvey was among the first to research a "dead" lake. Public awareness of acid rain in the U.S increased in the 1970s after The New York Times published reports from the Hubbard Brook Experimental Forest in New Hampshire of the myriad deleterious environmental effects shown to result from it.

Occasional pH readings in rain and fog water of well below 2.4 have been reported in industrialized areas. Industrial acid rain is a substantial problem in China and Russia and areas downwind from them. These areas all burn sulfur-containing coal to generate heat and electricity.

The problem of acid rain has not only increased with population and industrial growth, but has become more widespread. The use of tall smokestacks to reduce local pollution has contributed to the spread of acid rain by releasing gases into regional atmospheric circulation. Often deposition occurs a considerable distance downwind of the emissions, with mountainous regions tending to receive the greatest deposition (simply because of their higher rainfall). An example of this effect is the low pH of rain which falls in Scandinavia.

History of Acid Rain in the United States

In 1980, the U.S. Congress passed an Acid Deposition Act. This Act established an 18-year assessment and research program under the direction of the National Acidic Precipitation Assessment Program (NAPAP). NAPAP looked at the entire problem from a scientific perspective. It enlarged a network of monitoring sites to determine how acidic the precipitation actually was, and to determine long-term trends, and established a network for dry deposition. It looked at the effects of acid rain and funded research on the effects of acid precipitation on freshwater and terrestrial ecosystems, historical buildings, monuments, and building materials. It also funded extensive studies on atmospheric processes and potential control programs.

Since 1998, Harvard University wraps some of the bronze and marble statues on its campus, such as this "Chinese stele", with waterproof covers every winter, in order to protect them from erosion caused by acid rain and acid snow

From the start, policy advocates from all sides attempted to influence NAPAP activities to support their particular policy advocacy efforts, or to disparage those of their opponents. For the U.S. Government's scientific enterprise, a significant impact of NAPAP were lessons learned in the assessment process and in environmental research management to a relatively large group of scientists, program managers and the public.

In 1991, DENR provided its first assessment of acid rain in the United States. It reported that 5% of New England Lakes were acidic, with sulfates being the most common problem. They noted that 2% of the lakes could no longer support Brook Trout, and 6% of the lakes were unsuitable for the survival of many species of minnow. Subsequent Reports to Congress have documented chemical changes in soil and freshwater ecosystems, nitrogen saturation, decreases in amounts of nutrients in soil, episodic acidification, regional haze, and damage to historical monuments.

Meanwhile, in 1989, the U.S. Congress passed a series of amendments to the Clean Air Act. Title IV of these amendments established the Acid Rain Program, a cap and trade system designed to control emissions of sulfur dioxide and nitrogen oxides. Title IV called for a total reduction of about 10 million tons of SO_2 emissions from power plants. It was implemented in two phases. Phase I began in 1995, and limited sulfur dioxide emissions from 110 of the largest power plants to a combined total of 8.7 million tons of sulfur dioxide. One power plant in New England (Merrimack) was in Phase I. Four other plants (Newington, Mount Tom, Brayton Point, and Salem Harbor) were added under other provisions of the program. Phase II began in 2000, and affects most of the power plants in the country.

During the 1990s, research continued. On March 10, 2005, EPA issued the Clean Air Interstate Rule (CAIR). This rule provides states with a solution to the problem of power plant pollution that drifts from one state to another. CAIR will permanently cap emissions of SO_2 and NO_x in the eastern United States. When fully implemented, CAIR will reduce SO_2 emissions in 28 eastern states and the District of Columbia by over 70% and NO_x emissions by over 60% from 2003 levels.

Overall, the program's cap and trade program has been successful in achieving its goals. Since the 1990s, SO_2 emissions have dropped 40%, and according to the Pacific Research Institute, acid rain levels have dropped 65% since 1976. Conventional regulation was used in the European Union, which saw a decrease of over 70% in SO_2 emissions during the same time period.

In 2007, total SO_2 emissions were 8.9 million tons, achieving the program's long-term goal ahead of the 2010 statutory deadline.

In 2007 the EPA estimated that by 2010, the overall costs of complying with the program for businesses and consumers would be $1 billion to $2 billion a year, only one fourth of what was originally predicted. Forbes says: *In 2010, by which time the cap and trade system had been augmented by the George W. Bush administration's Clean Air Interstate Rule, SO2 emissions had fallen to 5.1 million tons.*

Emissions of Chemicals Leading to Acidification

The most important gas which leads to acidification is sulfur dioxide. Emissions of nitrogen oxides which are oxidized to form nitric acid are of increasing importance due to stricter controls on emissions of sulfur containing compounds. 70 Tg(S) per year in the form of SO_2 comes from fossil fuel combustion and industry, 2.8 Tg(S) from wildfires and 7–8 Tg(S) per year from volcanoes.

Natural Phenomena

The principal natural phenomena that contribute acid-producing gases to the atmosphere are emissions from volcanoes. Thus, for example, fumaroles from the Laguna Caliente crater of Poás Volcano create extremely high amounts of acid rain and fog, with acidity as high as a pH of 2, clearing an area of any vegetation and frequently causing irritation to the eyes and lungs of inhabitants in nearby settlements. Acid-producing gasses are also created by biological processes that occur on the land, in wetlands, and in the oceans. The major biological source of sulfur containing compounds is dimethyl sulfide.

Nitric acid in rainwater is an important source of fixed nitrogen for plant life, and is also produced by electrical activity in the atmosphere such as lightning.

Acidic deposits have been detected in glacial ice thousands of years old in remote parts of the globe.

Soils of coniferous forests are naturally very acidic due to the shedding of needles, and the results of this phenomenon should not be confused with acid rain.

Human Activity

The principal cause of acid rain is sulfur and nitrogen compounds from human sources, such as electricity generation, factories, and motor vehicles. Electrical power generation using coal is among the greatest contributors to gaseous pollutions that are responsible for acidic rain. The gases can be carried hundreds of kilometers in the atmosphere before they are converted to acids and deposited. In the past, factories had short funnels to let out smoke but this caused many problems locally; thus, factories now have taller smoke funnels. However, dispersal from these taller stacks causes pollutants to be carried farther, causing widespread ecological damage.

The coal-fired Gavin Power Plant in Cheshire, Ohio

Chemical Processes

Combustion of fuels produces sulfur dioxide and nitric oxides. They are converted into sulfuric acid and nitric acid.

Gas Phase Chemistry

In the gas phase sulfur dioxide is oxidized by reaction with the hydroxyl radical via an intermolecular reaction:

$$SO_2 + OH\cdot \rightarrow HOSO_2\cdot$$

which is followed by:

$$HOSO_2\cdot + O_2 \rightarrow HO_2\cdot + SO_3$$

In the presence of water, sulfur trioxide (SO_3) is converted rapidly to sulfuric acid:

$$SO_3\,(g) + H_2O\,(l) \rightarrow H_2SO_4\,(aq)$$

Nitrogen dioxide reacts with OH to form nitric acid:

$$NO_2 + OH\cdot \rightarrow HNO_3$$

Chemistry in Cloud Droplets

When clouds are present, the loss rate of SO_2 is faster than can be explained by gas phase chemistry alone. This is due to reactions in the liquid water droplets.

Hydrolysis

Sulfur dioxide dissolves in water and then, like carbon dioxide, hydrolyses in a series of equilibrium reactions:

$$SO_2\,(g) + H_2O \rightleftharpoons SO_2{\cdot}H_2O$$

$$SO_2{\cdot}H_2O \rightleftharpoons H^+ + HSO_3^-$$

$$HSO_3^- \rightleftharpoons H^+ + SO_3^{2-}$$

Oxidation

There are a large number of aqueous reactions that oxidize sulfur from S(IV) to S(VI), leading to the formation of sulfuric acid. The most important oxidation reactions are with ozone, hydrogen peroxide and oxygen (reactions with oxygen are catalyzed by iron and manganese in the cloud droplets).

Acid Deposition

Wet Deposition

Wet deposition of acids occurs when any form of precipitation (rain, snow, and so on.) removes acids from the atmosphere and delivers it to the Earth's surface. This can result from the deposition of acids produced in the raindrops or by the precipitation removing the acids either in clouds or below clouds. Wet removal of both gases and aerosols are both of importance for wet deposition.

Dry Deposition

Acid deposition also occurs via dry deposition in the absence of precipitation. This can be responsible for as much as 20 to 60% of total acid deposition. This occurs when particles and gases stick to the ground, plants or other surfaces.

Adverse Effects

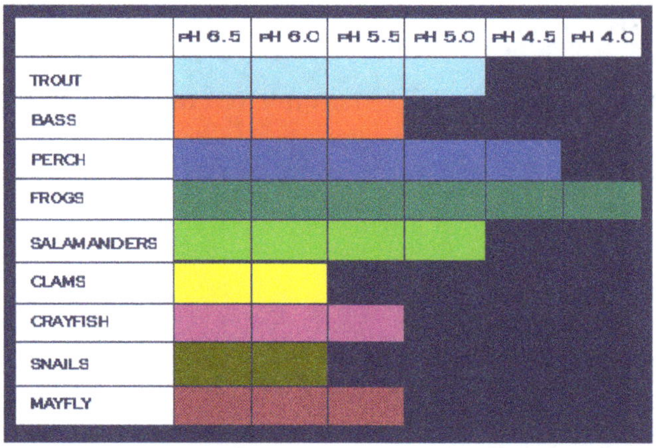

	pH 6.5	pH 6.0	pH 5.5	pH 5.0	pH 4.5	pH 4.0
TROUT						
BASS						
PERCH						
FROGS						
SALAMANDERS						
CLAMS						
CRAYFISH						
SNAILS						
MAYFLY						

This chart shows that not all fish, shellfish, or the insects that they eat can tolerate the same amount of acid; for example, frogs can tolerate water that is more acidic (i.e., has a lower pH) than trout.

Acid rain has been shown to have adverse impacts on forests, freshwaters and soils, killing insect and aquatic life-forms as well as causing damage to buildings and having impacts on human health.

Surface Waters and Aquatic Animals

Both the lower pH and higher aluminium concentrations in surface water that occur as a result of acid rain can cause damage to fish and other aquatic animals. At pHs lower than 5 most fish eggs will not hatch and lower pHs can kill adult fish. As lakes and rivers become more acidic biodiversity is reduced. Acid rain has eliminated insect life and some fish species, including the brook trout in some lakes, streams, and creeks in geographically sensitive areas, such as the Adirondack Mountains of the United States. However, the extent to which acid rain contributes directly or indirectly via runoff from the catchment to lake and river acidity (i.e., depending on characteristics of the surrounding watershed) is variable. The United States Environmental Protection Agency's (EPA) website states: "Of the lakes and streams surveyed, acid rain caused acidity in 75% of the acidic lakes and about 50% of the acidic streams".

Soils

Soil biology and chemistry can be seriously damaged by acid rain. Some microbes are unable to tolerate changes to low pH and are killed. The enzymes of these microbes are denatured (changed in shape so they no longer function) by the acid. The hydronium ions of acid rain also mobilize toxins such as aluminium, and leach away essential nutrients and minerals such as magnesium.

$$2\,H^+\,(aq) + Mg^{2+}\,(clay) \rightleftharpoons 2\,H^+\,(clay) + Mg^{2+}\,(aq)$$

Soil chemistry can be dramatically changed when base cations, such as calcium and magnesium, are leached by acid rain thereby affecting sensitive species, such as sugar maple (Acer saccharum).

Forests and Other Vegetation

Adverse effects may be indirectly related to acid rain, like the acid's effects on soil or high concentration of gaseous precursors to acid rain. High altitude forests are especially vulnerable as they are often surrounded by clouds and fog which are more acidic than rain.

Other plants can also be damaged by acid rain, but the effect on food crops is minimized by the application of lime and fertilizers to replace lost nutrients. In cultivated areas, limestone may also be added to increase the ability of the soil to keep the pH stable, but this tactic is largely unusable in the case of wilderness lands. When calcium is leached from the needles of red spruce, these trees become less cold tolerant and exhibit winter injury and even death.

Ocean Acidification

Coral's limestone skeletal is sensitive to pH drop, because the calcium carbonate, core component of the limestone dissolves in acidic (low pH) solutions.

Human Health Effects

Acid rain does not directly affect human health. The acid in the rainwater is too dilute to have direct adverse effects. However, the particulates responsible for acid rain (sulfur dioxide and nitrogen oxides) do have an adverse effect. Increased amounts of fine particulate matter in the air do contribute to heart and lung problems including asthma and bronchitis.

Other Adverse Effects

Effect of acid rain on statues

Acid rain can damage buildings, historic monuments, and statues, especially those made of rocks, such as limestone and marble, that contain large amounts of calcium carbonate. Acids in the rain react with the calcium compounds in the stones to create gypsum, which then flakes off.

$$CaCO_3\,(s) + H_2SO_4\,(aq) \rightleftharpoons CaSO_4\,(s) + CO_2\,(g) + H_2O\,(l)$$

Acid rain and weathering

The effects of this are commonly seen on old gravestones, where acid rain can cause the inscriptions to become completely illegible. Acid rain also increases the corrosion rate of metals, in particular iron, steel, copper and bronze.

Affected Areas

Places significantly impacted by acid rain around the globe include most of eastern Europe from Poland northward into Scandinavia, the eastern third of the United States, and southeastern Canada. Other affected areas include the southeastern coast of China and Taiwan.

Prevention Methods

Technical Solutions

Many coal-firing power stations use flue-gas desulfurization (FGD) to remove sulfur-containing gases from their stack gases. For a typical coal-fired power station, FGD will remove 95% or more of the SO_2 in the flue gases. An example of FGD is the wet scrubber which is commonly used. A wet scrubber is basically a reaction tower equipped with a fan that extracts hot smoke stack gases from a power plant into the tower. Lime or limestone in slurry form is also injected into the tower to mix with the stack gases and combine with the sulfur dioxide present. The calcium carbonate of the limestone produces pH-neutral calcium sulfate that is physically removed from the scrubber. That is, the scrubber turns sulfur pollution into industrial sulfates.

In some areas the sulfates are sold to chemical companies as gypsum when the purity of calcium sulfate is high. In others, they are placed in landfill. However, the effects of acid rain can last for generations, as the effects of pH level change can stimulate the continued leaching of undesirable chemicals into otherwise pristine water sources, killing off vulnerable insect and fish species and blocking efforts to restore native life.

Fluidized bed combustion also reduces the amount of sulfur emitted by power production.

Vehicle emissions control reduces emissions of nitrogen oxides from motor vehicles.

International Treaties

A number of international treaties on the long-range transport of atmospheric pollutants have been agreed for example, the 1985 Helsinki Protocol on the Reduction of Sulphur Emissions under the Convention on Long-Range Transboundary Air Pollution. Canada and the US signed the Air Quality Agreement in 1991. Most European countries and Canada have signed the treaties.

Emissions Trading

In this regulatory scheme, every current polluting facility is given or may purchase on an open market an emissions allowance for each unit of a designated pollutant it emits. Operators can then install pollution control equipment, and sell portions of their emissions allowances they no longer need for their own operations, thereby recovering some of the capital cost of their investment in such equipment. The intention is to give operators economic incentives to install pollution controls.

The first emissions trading market was established in the United States by enactment of the Clean Air Act Amendments of 1990. The overall goal of the Acid Rain Program established by the Act is to achieve significant environmental and public health benefits through reductions in emissions of sulfur dioxide (SO_2) and nitrogen oxides (NO_x), the primary causes of acid rain. To achieve this goal at the lowest cost to society, the program employs both regulatory and market based approaches for controlling air pollution.

Air Quality Health Index (Canada)

The Air Quality Health Index or (AQHI) is a scale designed in Canada to help understand the impact of air quality on health. It is a health protection tool used to make decisions to reduce short-term exposure to air pollution by adjusting activity levels during increased levels of air pollution. The Air Quality Health Index also provides advice on how to improve air quality by proposing behavioral change to reduce the environmental footprint. This index pays particular attention to people who are sensitive to air pollution. It provides them with advice on how to protect their health during air quality levels associated with low, moderate, high and very high health risks.

History

Air quality in Canada has been historically been reported by the USA's Air Quality Index in various provinces. Significantly, AQI values reflect air quality management objectives, which are based on the lowest achievable emissions rate, and not exclusively concern for human health. The AQHI was created with a different goal - to report on the specific health risks posed by air pollution. As such, the AQHI represents a paradigm shift in communicating air quality information to the public.

The Air Quality Health Index or "AQHI" is a federal program jointly coordinated by Health Canada and Environment Canada. However, the AQHI program would not be possible without the commitment and support of the provinces, municipalities and NGOs. From air quality monitoring to health risk communication and community engagement, local partners are responsible for the vast majority of work related to AQHI implementation.

The AQHI is being rolled out across Canada and replacing the AQI as the public face of air quality information.

Originally launched as a pilot project in the British Columbia Interior and Nova Scotia in 2005, it is currently implemented in 79 locations across Canada.

Overview

The *Air Quality Health Index* provides a number from 1 to 10+ to indicate the level of health risk associated with local air quality. Occasionally, when the amount of air pollution is abnormally high, the number may exceed 10. The AQHI provides a local air quality current value as well as a local air quality maximums forecast for today, tonight and tomorrow and provides associated health advice.

Risk: Low (1-3) Moderate (4-6) High (7-10) Very high (above 10)

As it is now known that even low levels of air pollution can trigger discomfort for the sensitive population, the index has been developed as a continuum: The higher the number, the greater the health risk and need to take precautions. The index describes the level of health risk associated with this number as 'low', 'moderate', 'high' or 'very high', and suggests steps that can be taken to reduce exposure.

Health Risk	Air Quality Health Index	Health Messages for At-Risk Population	Health messages for General Population
Low	1-3	**Enjoy** your usual outdoor activities.	**Ideal** air quality for outdoor activities
Moderate	4-6	**Consider reducing** or rescheduling strenuous activities outdoors if you are experiencing symptoms.	**No need to modify** your usual outdoor activities unless you experience symptoms such as coughing and throat irritation.
High	7-10	**Reduce** or reschedule strenuous activities outdoors. Children and the elderly should also take it easy.	**Consider reducing** or rescheduling strenuous activities outdoors if you experience symptoms such as coughing and throat irritation.
Very high	Above 10	**Avoid** strenuous activities outdoors. Children and the elderly should also avoid outdoor physical exertion.	**Reduce** or reschedule strenuous activities outdoors, especially if you experience symptoms such as coughing and throat irritation.

Formula Inputs

The formulation of the national AQHI is based on the observed relationship of nitrogen dioxide

(NO_2), ground-level ozone (O_3) and fine particulate matter ($PM_{2.5}$) with mortality from an analysis of several Canadian cities. Significantly, all three of these pollutants can pose health risks, even at low levels of exposure, especially among those with pre-existing health problems.

When developing the AQHI, Health Canada's original analysis of health effects included five major air pollutants: airborne particulate matter, ozone, and nitrogen dioxide (NO_2), as well as sulphur dioxide (SO_2), and carbon monoxide (CO). The latter two pollutants provided little information in predicting health effects and were removed from the AQHI formulation.

The AQHI does not measure the effects of odour, pollen, dust, heat or humidity.

Calculation

The national AQHI is based on three-hour average concentrations of ground-level ozone (O_3), nitrogen dioxide (NO_2), and fine particulate matter (PM2.5). O_3 and NO_2 are measured in parts per billion (ppb) while PM2.5 is measured in micrograms per cubic metre (ug/m³).

The AQHI is calculated on a community basis (each community may have one or more monitoring stations).

First, the average concentration of the 3 substances (O_3, NO_2, PM2.5) is calculated at each station within a community for the 3 preceding hours. This is considered valid only if at least 2 out of 3 hours are available at the station. If more than 1 of the preceding 3 hours is missing the station average is set to "Not Available". This part of the process results in three "station parameter averages" for each station.

Second, the 3 hour "community average" for each parameter is calculated from the 3 hour substance averages at the available stations. If no stations are available for a parameter, that parameter is set to "Not Available". This part of the process results in 3 community parameter averages.

Third, if all three community parameter averages are available, a community AQHI is calculated. The formula is:

$$AQHI = (\frac{1000}{10.4}) \times [(e^{0.000537 \times O_3} - 1) + (e^{0.000871 \times NO_2} - 1) + (e^{0.000487 \times PM_{2.5}} - 1)]$$

The result is then rounded to the nearest positive integer; a calculation less than 0.5 is rounded up to 1.

Alberta Calculation and Reporting Differences

Alberta has modified AQHI reporting to better suit the needs of the Province. Because of Alberta's energy based economy other are also considered when reporting the AQHI.

Alberta also has rapidly changingair quality conditions quite often (for example during wildfire season) so, Alberta's AQHI needs to be more responsive than the national AQHI, which is based on a three-hour average.

In order to meet these needs, the individual pollutant concentrations are compared to Alberta's

Ambient Air Quality Objectives (AAQOs). The national AQHI is used most of the time; however, if hourly air pollutant concentrations are higher than Alberta's AAQOs, the AQHI value is replaced (overridden) with the appropriate "High" or "Very High" risk value. This can occur for the following pollutants (when they exceed the noted concentrations):

- 80 micrograms per cubic metre for fine particulate matter
- 82 parts per billion for ozone
- 159 parts per billion for nitrogen dioxide
- 172 parts per billion for sulphur dioxide
- 13 parts per million for carbon monoxide
- 1 part per million for hydrogen sulphide and total reduced sulphur

In Alberta, special community messaging is used when the level of specific pollutants is higher than specified odour or visibility thresholds but the AQHI is rated as "Low" or "Moderate" risk. This messaging is used for the following pollutants (when they exceed the noted concentrations):

- 25 micrograms per cubic metre for fine particulate matter (based on visibility)
- 100 parts per billion for sulphur dioxide (based on odour)
- 10 parts per billion for hydrogen sulphide or total reduced sulphur (based on odour)

This odour/visibility messaging appears as below:

Alberta has developed its own AQHI website at http://airquality.alberta.ca.

Persons at Risk

The *AQHI* is aimed towards two populations: 1. The "general" population; and 2. The "at-risk" populations. The later consists of children, the elderly and people with existing respiratory or cardiovascular conditions, such as those with asthma, and people suffering from diabetes, heart disease or lung disease.

Children are more vulnerable to air pollution: they have less-developed respiratory and defense systems. Because of their size, they inhale more air per kilogram of body weight than adults. Their elevated metabolic rate and young defense systems make them more susceptible to air pollution.

Seniors are also at a higher risk because of the weakening of the heart, lungs and immune system and increased likelihood of health problems such as heart and lung disease.

Exposure to air pollutants can cause a range of symptoms. People with lung or heart disease may experience increased frequency and/or severity of symptoms, and increased medication requirements. It is recommended that those susceptive should take greater precautions.

Lifestyle

Environment Canada recommends looking for outdoor air quality by checking the *AQHI* in your community before heading off to work or play as well as to use the forecasts to plan activities, whether over the next hour or the next day. Seniors, parents, those with asthma, and people suffering from diabetes, heart or lung disease, can use the *AQHI* to assess the immediate risk air pollution poses on their health and take steps to lessen that risk. The *AQHI* is also recommended for healthy, fit and active people to consult to decide when it is best to exercise or work outdoor.

The best way for someone to use the *AQHI* is to regularly check the current index value, to pay attention to personal symptoms and self-calibrate to the reported current *AQHI* value. For example, if symptoms are experienced when the index is a 6, then precaution should be taken when the index is at a 6 or higher by following the corresponding health messages. Then, when an individual knows what number triggers health symptoms, to get in the habit of checking the maximum forecast to plan activities ahead of time.

The *AQHI* is easily accessible via the web: At http://www.airhealth.ca and http://www.weatheroffice.gc.ca/canada_e.html and The Weather Network. It is reported throughout the day on The Weather Network channel as well as the weather and traffic reports on local media.The AQHI is also available for download as a desktop widget for Windows and Mac, and iPhone app, http://mirtchovski.com/code/AQHI.html. There are several news feeds and blogs dedicated to the AQHI: http://www.ec.gc.ca/cas-aqhi/default.asp?lang=En&n=C7B2359F-1.

Air Quality Law

Air quality laws govern the emission of air pollutants into the atmosphere. A specialized subset of air quality laws regulate the quality of air inside buildings. Air quality laws are often designed specifically to protect human health by limiting or eliminating airborne pollutant concentrations. Other initiatives are designed to address broader ecological problems, such as limitations on chemicals that affect the ozone layer, and emissions trading programs to address acid rain or climate change. Regulatory efforts include identifying and categorizing air pollutants, setting limits on acceptable emissions levels, and dictating necessary or appropriate mitigation technologies.

Air Pollutant Classification

Air quality regulation must identify the substances and energies which qualify as "pollution" for purposes of further control. While specific labels vary from jurisdiction to jurisdiction, there is broad consensus among many governments regarding what constitutes air pollution. For example, the United States Clean Air Act identifies ozone, particulate matter, carbon monoxide, nitrogen oxides (NO_x), sulfur dioxide (SO_2), and lead (Pb) as "criteria" pollutants requiring nationwide regulation. EPA has also identified over 180 compounds it has classified as "hazardous" pollutants requiring strict control. Other compounds have been identified as air pollutants due to their adverse impact on the environment (e.g., CFCs as agents of ozone depletion), and on human health (e.g., asbestos in indoor air). A broader conception of air pollution may also incorporate noise, light, and radiation. The United States has recently seen controversy over whether carbon dioxide (CO_2) and other greenhouse gases should be classified as air pollutants.

Air Quality Standards

Air quality standards are legal standards or requirements governing concentrations of air pollutants in breathed air, both outdoors and indoors. Such standards generally are expressed as levels of specific air pollutants that are deemed acceptable in ambient air, and are most often designed to reduce or eliminate the human health effects of air pollution, although secondary effects such as crop and building damage may also be considered. Determining appropriate air quality standards generally requires up-to-date scientific data on the health effects of the pollutant under review, with specific information on exposure times and sensitive populations. It also generally requires periodic or continuous monitoring of air quality.

As an example, the United States Environmental Protection Agency has developed the National Ambient Air Quality Standards (NAAQS) NAAQS set attainment thresholds for sulfur dioxide, particulate matter (PM_{10} and $PM_{2.5}$), carbon monoxide, ozone, nitrogen oxides NO_x, and lead (Pb) in outdoor air throughout the United States. Another set of standards, for indoor air in employment settings, is administered by the U.S. Occupational Safety and Health Administration.

A distinction may be made between mandatory and aspirational air quality standards. For example, U.S. state governments must work toward achieving NAAQS, but are not forced to meet them. On the other hand, employers may be required immediately to rectify any violation of OSHA workplace air quality standards.

Emission Standards

Emission standards are the legal requirements governing air pollutants released into the atmosphere. Emission standards set quantitative limits on the permissible amount of specific air pollutants that may be released from specific sources over specific timeframes. They are generally designed to achieve air quality standards and to protect human health.

Numerous methods exist for determining appropriate emissions standards, and different regulatory approaches may be taken depending on the source, industry, and air pollutant under review. Specific limits may be set by reference to and within the confines of more general air quality standards. Specific sources may be regulated by means of performance standards, meaning numerical limits on the emission of a specific pollutant from that source category. Regulators may also mandate the adoption and use of specific control technologies, often with reference to feasibility, availability, and cost. Still other standards may be set using performance as a benchmark - for example, requiring all of a specific type of facility to meet the emissions limits achieved by the best performing facility of the group. All of these methods may be modified by incorporating emissions averaging, market mechanisms such as emissions trading, and other alternatives.

For example, all of these approaches are used in the United States. The United States Environmental Protection Agency (responsible for air quality regulation at a national level under the U.S. Clean Air Act, utilizes performance standards under the New Source Performance Standard (NSPS) program. Technology requirements are set under RACT (Reasonably Available Control Technology), BACT (Best Available Control Technology), and LAER (Lowest Achievable Emission Rate) standards. Flexibility alternatives are implemented in U.S. programs to eliminate acid rain, protect the ozone layer, achieve permitting standards, and reduce greenhouse gas emissions.

Control Technology Requirements

In place of or in combination with air quality standards and emission control standards, governments may choose to reduce air pollution by requiring regulated parties to adopt emissions control technologies (i.e., technology that reduces or eliminates emissions). Such devices include but are not limited to flare stacks, incinerators, catalytic combustion reactors, selective catalytic reduction reactors, electrostatic precipitators, baghouses, wet scrubbers, cyclones, thermal oxidizers, Venturi scrubbers, carbon adsorbers, and biofilters.

The selection of emissions control technology may be the subject of complex regulation that may balance multiple conflicting considerations and interests, including economic cost, availability, feasibility, and effectiveness. The various weight given to each factor may ultimately determine the technology selected. The outcome of an analysis seeking a technology that all players in an industry can afford could be different from an analysis seeking to require all players to adopt the most effective technology yet developed, regardless of cost. For example, the United States Clean Air Act contains several control technology requirements, including Best Available Control Technology (BACT) (used in New Source Review), Reasonably Available Control Technology (RACT) (existing sources), Lowest Achievable Emissions Rate (LAER) (used for major new sources in non-attainment areas), and Maximum Achievable Control Technology (MACT) standards.

Bans

Air quality laws may take the form of bans. While arguably a class of emissions control law (where the emission limit is set to zero), bans differ in that they may regulate activity other than the emission of a pollutant itself, even though the ultimate goal is to eliminate the emission of the pollutant.

A common example is a burn ban. Residential and commercial burning of wood materials may be restricted during times of poor air quality, eliminating the immediate emission of particulate matter and requiring use of non-polluting heating methods. A more significant example is the widespread ban on the manufacture of dichlorodifluoromethane (Freon)), formerly the standard refrigerant in automobile air conditioning systems. This substance, often released into the atmosphere unintentionally as a result of refrigerant system leaks, was determined to have a significant ozone depletion potential, and its widespread use to pose a significant threat to the Earth's ozone layer. Its manufacture was prohibited as part of a suite of restrictions adopted internationally in the Montreal Protocol to the Vienna Convention for the Protection of the Ozone Layer. Still another example is the ban on use of asbestos in building construction materials, to eliminate future exposure to carcinogenic asbestos fibers when the building materials are disturbed.

Data Collection and Access

Air quality laws may impose substantial requirements for collecting, storing, submitting, and providing access to technical data for various purposes, including regulatory enforcement, public health programs, and policy development.

Data collection processes may include monitoring ambient air for the presence of pollutants, directly monitoring emissions sources, or collecting other quantitative information from which air quality information may be deduced. For example, local agencies may employ a particulate matter sampler to determine ambient air quality in a locality over time. Fossil power plants may required to monitor emissions at a flue-gas stack to determine quantities of relevant pollutants emitted. Automobile manufacturers may be required to collect data regarding car sales, which, when combined with technical specifications regarding fuel consumption and efficiency, may be used to estimate total vehicle emissions. In each case, data collection may be short- or long-term, and at varying frequency (e.g., hourly, daily).

Air quality laws may include detailed requirements for recording, storing, and submitting relevant information, generally with the ultimate goal of standardizing data practices in order to facilitate data access and manipulation at a later time. Precise requirements may be very difficult to determine without technical training and may change over time in response to, for example, changes in law, changes in policy, changes in available technology, and changes in industry practice. Such requirements may be developed at a national level and reflect consensus or compromise between government agencies, regulated industry, and public interest groups.

Once air quality data are collected and submitted, some air quality laws may require government agencies or private parties to provide the public with access to the information - whether the raw data alone, or via tools to make the data more useful, accessible, and understandable. Where public access mandates are general, it may be left to the collecting agency to decide whether and to what extent the data is to be centralized and organized. For example, the United States Environmental Protection Agency, National Oceanic and Atmospheric Administration, National Park Service, and tribal, state, and local agencies coordinate to produce an online mapping and data access tool called AirNow, which provides real-time public access to U.S. air quality index information, searchable by location.

Once data are collected and published, they may be used as inputs in mathematical models and forecasts. For example, atmospheric dispersion modeling may be employed to examine the potential impact of new regulatory requirements on existing populations or geographic areas. Such models in turn could drive changes in data collection and reporting requirements.

Controversy

Proponents of air quality law argue that they have caused or contributed to major reductions in air pollution, with concomitant human health and environmental benefits, even in the face of large-scale economic growth and increases in motor vehicle use. On the other hand, controversy may arise over the estimated cost of additional regulatory standards.

Arguments over cost, however, cut both ways. For example, the "estimates that the benefits of reducing fine particle and ground level ozone pollution under the 1990 Clean Air Act amendments will reach approximately $2 trillion in 2020 while saving 230,000 people from early death in that year alone." According to the same report, 2010 alone the reduction of ozone and particulate matter in the atmosphere prevented more than 160,000 cases of premature mortality, 130,000 heart attacks, 13 million lost work days and 1.7 million asthma attacks. Criticisms of EPA's methodologies in reaching these and similar numbers are publicly available.

Around the World

International Law

International law includes agreements related to trans-national air quality, including greenhouse gas emissions:

- Convention on Long-Range Transboundary Air Pollution (LRTAP), Geneva, 1979.
- Environmental Protection: Aircraft Engine Emissions, Annex 16, vol. 2 to the Chicago Convention on International Civil Aviation, Montreal, 1981.
- Framework Convention on Climate Change (UNFCCC), New York, 1992, including the Kyoto Protocol, 1997, and the Paris Agreement, 2015.
- Georgia Basin-Puget Sound International Airshed Strategy, Vancouver, Statement of Intent, 2002.
- Vienna Convention for the Protection of the Ozone Layer, Vienna, 1985, including the Montreal Protocol on Substances that Deplete the Ozone Layer, Montreal 1987.
- U.S.-Canada Air Quality Agreement (bilateral U.S.-Canadian agreement on acid rain), 1986

Canada

With some industry-specific exceptions, Canadian air pollution regulation was traditionally handled at the provincial level. However, under the authority of the Canadian Environmental Protection Act, 1999, the country has recently enacted a national program called the Canadian Air Quality Management System (AQMS). The program includes five main regulatory mechanisms: the Canadian Ambient Air Quality Standards (CAAQS); Base Level Industrial Emission Requirements (BLIERs) (emissions controls and technology); management of local air quality through the management of Local Air Zones; management of regional air quality through the management of Regional Airsheds; and collaboration to reduce mobile source emissions.

The Canadian government has also made efforts to pass legislation related to the country's greenhouse gas emissions. It has passed laws related to fuel economy in passenger vehicles and light trucks, heavy-duty vehicles, renewable fuels, and the energy and transportation sectors.

China

China, with severe air pollution in mega-cities and industrial centers, particularly in the north, has adapted the Airborne Pollution Prevention and Control Action Plan which aims for a 25% reduction in air pollution by 2017 from 2012 levels. Funded by $277 billion from the central government, the action plan targets PM 2.5 particulates which affect human health.

New Zealand

New Zealand passed its Clean Air Act 1972 in response to growing concerns over industrial and urban air pollution. That Act classified sources, imposed permitting requirements, and created a process for determining requisite control technology. Local authorities were authorized to regulate smaller polluters. Within the Christchurch Clean Air Zone, burn bans and other measures were effected to control smog.

The Clean Air Act 1972 was replaced by the Resource Management Act 1991. The act did not set air quality standards, but did provide for national guidance to be developed. This resulted in the promulgation of New Zealand's National Environmental Standards for Air Quality in 2004 with subsequent amendments.

United Kingdom

In response to the Great Smog of 1952, the British Parliament introduced the Clean Air Act 1956. This act legislated for zones where smokeless fuels had to be burnt and relocated power stations to rural areas. The Clean Air Act 1968 introduced the use of tall chimneys to disperse air pollution for industries burning coal, liquid or gaseous fuels. The Clean Air Act was updated in 1993 and can be reviewed online legislation Clean Air Act 1993. The biggest domestic impact comes from Part III, Smoke Control Areas, which are designated by local authorities and can vary by street in large towns.

United States

The primary law regulating air quality in the United States is the U.S. Clean Air Act. The law was initially enacted as the Air Pollution Control Act of 1955. Amendments in 1967 and 1970 (the framework for today's U.S. Clean Air Act) imposed national air quality requirements, and placed administrative responsibility with the newly created Environmental Protection Agency. Major amendments followed in 1977 and 1990. State and local governments have enacted similar legislation, either implementing federal programs or filling in locally important gaps in federal programs.

References

- Garcia, Javier; Colosio, Joëlle (2002). Air-quality indices : elaboration, uses and international comparisons. Presses des MINES. ISBN 2-911762-36-3.
- Weathers, K. C. and Likens, G. E. (2006). "Acid rain", pp. 1549–1561 in: W. N. Rom and S. Markowitz (eds.). Environmental and Occupational Medicine. Lippincott-Raven Publ., Philadelphia. Fourth Edition, ISBN 0-7817-6299-5.
- Seinfeld, John H.; Pandis, Spyros N (1998). Atmospheric Chemistry and Physics — From Air Pollution to Climate Change. John Wiley and Sons, Inc. ISBN 978-0-471-17816-3
- Berresheim, H.; Wine, P.H. and Davies D.D. (1995). "Sulfur in the Atmosphere". In Composition, Chemistry and Climate of the Atmosphere, ed. H.B. Singh. Van Nostrand Rheingold ISBN 0-442-01264-0
- DeHayes, D.H., Schaberg, P.G. and G.R. Strimbeck. (2001). Red Spruce Hardiness and Freezing Injury Susceptibility. In: F. Bigras, ed. Conifer Cold Hardiness. Kluwer Academic Publishers, the Netherlands ISBN 0-7923-6636-0.
- National Weather Service Corporate Image Web Team. "NOAA's National Weather Service/Environmental Protection Agency - United States Air Quality Forecast Guidance". Retrieved 20 August 2015.
- Rama Lakshmi (17 October 2014). "India launches its own Air Quality Index. Can its numbers be trusted?". Washington Post. Retrieved 20 August 2015.
- "National Air Quality Index (AQI) launched by the Environment Minister AQI is a huge initiative under 'Swa-

chh Bharat'". Retrieved 20 August 2015.

- Gerdes, Justin. "Cap and Trade Curbed Acid Rain: 7 Reasons Why It Can Do The Same For Climate Change". Forbes. Forbes. Retrieved 27 Oct 2014.
- "How is AQHI Reporting Enhanced in Alberta? – Alberta Environment and Sustainable Resource Development". Environment.alberta.ca. 2011-06-14. Retrieved 2013-07-23.
- David Mintz (February 2009). Technical Assistance Document for the Reporting of Daily Air Quality – the Air Quality Index (AQI) (PDF). North Carolina: US EPA Office of Air Quality Planning and Standards. EPA-454/B-09-001. Retrieved 9 August 2012.
- Enesta Jones (03/01/2011). "EPA Report Underscores Clean Air Act's Successful Public Health Protections/ Landmark law saved 160,000 lives in 2010 alone". EPA.gov. Retrieved 22 March 2012.
- Chandru (2006-09-09). "CHINA: Industrialization pollutes its country side with Acid Rain". Southasiaanalysis.org. Archived from the original on June 20, 2010. Retrieved 2010-11-18.

8

Devices to Combat Air Pollution

Devices and techniques are an important component of any field of study. The following chapter elucidates the various devices that can be used to fight air pollution. The ones introduced in this chapter are cyclonic separation, electrostatic precipitator, dust collector and wet scrubber. This chapter serves as a source to understand these devices better.

Cyclonic Separation

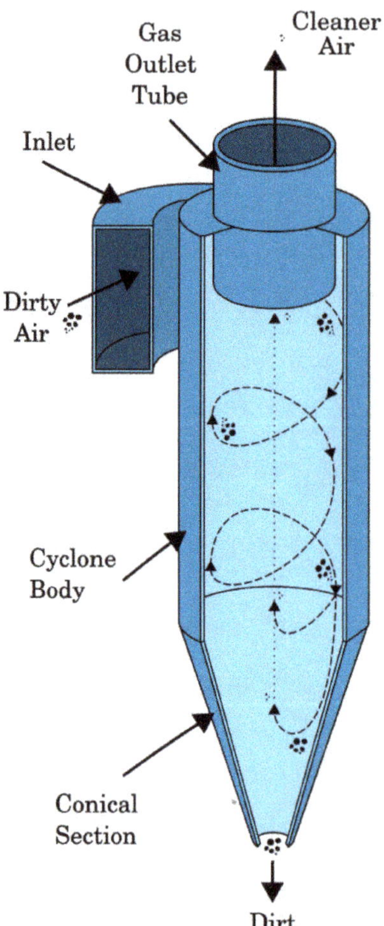

A simple cyclone separator

Cyclonic separation is a method of removing particulates from an air, gas or liquid stream, without the use of filters, through vortex separation. When removing particulate matter from liquids, a hydrocyclone is used; while from gas, a gas cyclone is used. Rotational effects and gravity are used to separate mixtures of solids and fluids. The method can also be used to separate fine droplets of liquid from a gaseous stream.

A high speed rotating (air)flow is established within a cylindrical or conical container called a cyclone. Air flows in a helical pattern, beginning at the top (wide end) of the cyclone and ending at the bottom (narrow) end before exiting the cyclone in a straight stream through the center of the cyclone and out the top. Larger (denser) particles in the rotating stream have too much inertia to follow the tight curve of the stream, and strike the outside wall, then fall to the bottom of the cyclone where they can be removed. In a conical system, as the rotating flow moves towards the narrow end of the cyclone, the rotational radius of the stream is reduced, thus separating smaller and smaller particles. The cyclone geometry, together with flow rate, defines the *cut point* of the cyclone. This is the size of particle that will be removed from the stream with a 50% efficiency. Particles larger than the cut point will be removed with a greater efficiency, and smaller particles with a lower efficiency.

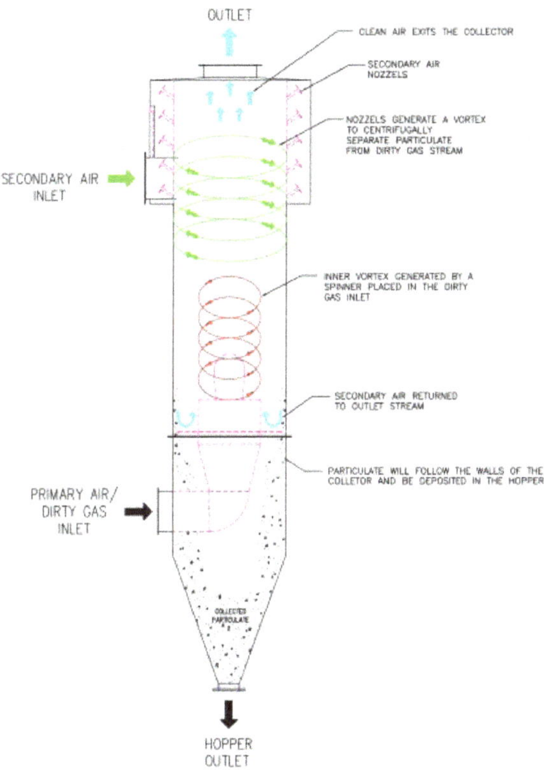

Airflow diagram for Aerodyne cyclone in standard vertical position. Secondary air flow is injected to reduce wall abrasion.

An alternative cyclone design uses a secondary air flow within the cyclone to keep the collected particles from striking the walls, to protect them from abrasion. The primary air flow containing the particulates enters from the bottom of the cyclone and is forced into spiral rotation by stationary spinner vanes. The secondary air flow enters from the top of the cyclone and moves downward toward the bottom, intercepting the particulate from the primary air. The secondary air flow also allows the collector to optionally be mounted horizontally, because it pushes the particulate toward the collection area, and does not rely solely on gravity to perform this function.

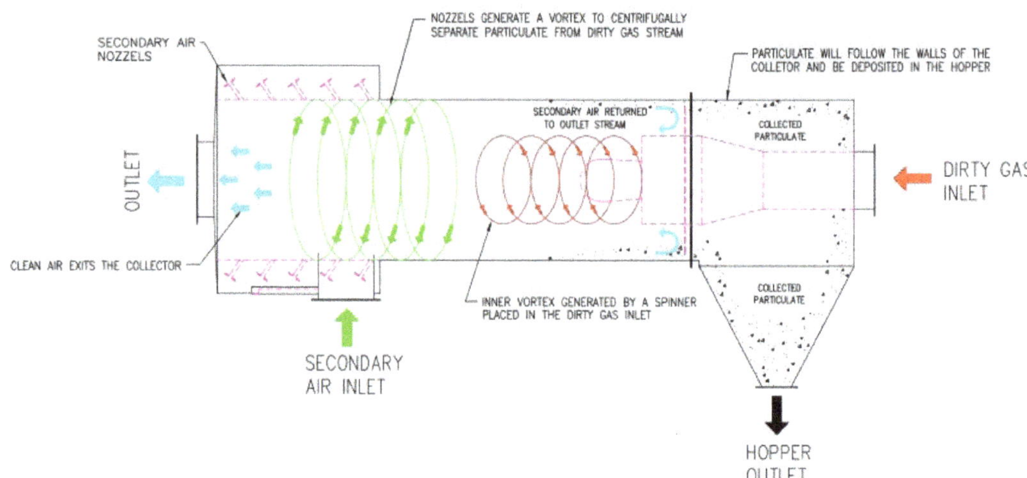

Airflow diagram for Aerodyne cyclone in horizontal position, an alternate design. Secondary air flow is injected to reduce wall abrasion, and to help move collected particulates to hopper for extraction.

Large scale cyclones are used in sawmills to remove sawdust from extracted air. Cyclones are also used in oil refineries to separate oils and gases, and in the cement industry as components of kiln preheaters. Cyclones are increasingly used in the household, as the core technology in bagless types of portable vacuum cleaners and central vacuum cleaners. Cyclones are also used in industrial and professional kitchen ventilation for separating the grease from the exhaust air in extraction hoods. Smaller cyclones are used to separate airborne particles for analysis. Some are small enough to be worn clipped to clothing, and are used to separate respirable particles for later analysis.

Similar separators are used in the oil refining industry (e.g. for Fluid catalytic cracking) to achieve fast separation of the catalyst particles from the reacting gases and vapors.

Analogous devices for separating particles or solids from liquids are called hydrocyclones or hydroclones. These may be used to separate solid waste from water in wastewater and sewage treatment.

Cyclone Theory

As the cyclone is essentially a two phase particle-fluid system, fluid mechanics and particle transport equations can be used to describe the behaviour of a cyclone. The air in a cyclone is initially introduced tangentially into the cyclone with an inlet velocity V_{in}. Assuming that the particle is spherical, a simple analysis to calculate critical separation particle sizes can be established.

If one considers an isolated particle circling in the upper cylindrical component of the cyclone at a rotational radius of r from the cyclone's central axis, the particle is therefore subjected to drag, centrifugal, and buoyant forces. Given that the fluid velocity is moving in a spiral the gas velocity can be broken into two component velocities: a tangential component, V_t, and an outward radial velocity component V_r. Assuming Stokes' law, the drag force in the outward radial direction that is opposing the outward velocity on any particle in the inlet stream is:

$$F_d = -6\pi r_p \mu V_r.$$

Using ρ_p as the particles density, the centrifugal component in the outward radial direction is:

$$F_c = m\frac{V_t^2}{r}$$

$$= \frac{4}{3}\pi\rho_p r_p^3 \frac{V_t^2}{r}.$$

The buoyant force component is in the inward radial direction. It is in the opposite direction to the particle's centrifugal force because it is on a volume of fluid that is missing compared to the surrounding fluid. Using ρ_f for the density of the fluid, the buoyant force is:

$$F_b = -V_p \rho_f \frac{V_t^2}{r}$$

$$= -\frac{4\pi r_p^3}{3}\frac{V_t^2}{r}\rho_f.$$

In this case, V_p is equal to the volume of the particle (as opposed to the velocity). Determining the outward radial motion of each particle is found by setting Newton's second law of motion equal to the sum of these forces:

$$m\frac{dV_r}{dt} = F_d + F_c + F_b$$

To simplify this, we can assume the particle under consideration has reached "terminal velocity", i.e., that its acceleration $\frac{dV_r}{dt}$ is zero. This occurs when the radial velocity has caused enough drag force to counter the centrifugal and buoyancy forces. This simplification changes our equation to:

$$F_d + F_c + F_b = 0$$

Which expands to:

$$-6\pi r_p \mu V_r + \frac{4}{3}\pi r_p^3 \frac{V_t^2}{r}\rho_p - \frac{4}{3}\pi r_p^3 \frac{V_t^2}{r}\rho_f = 0$$

Solving for V_r we have

$$V_r = \frac{2}{9}\frac{r_p^2}{\mu}\frac{V_t^2}{r}(\rho_p - \rho_f).$$

Notice that if the density of the fluid is greater than the density of the particle, the motion is (-), toward the center of rotation and if the particle is denser than the fluid, the motion is (+), away from the center. In most cases, this solution is used as guidance in designing a separator, while actual performance is evaluated and modified empirically.

In non-equilibrium conditions when radial acceleration is not zero, the general equation from above must be solved. Rearranging terms we obtain

$$\frac{dV_r}{dt}+\frac{9}{2}\frac{\mu}{\rho_p r_p^2}V_r-\left(1-\frac{\rho_f}{\rho_p}\right)\frac{V_t^2}{r}=0$$

Since V_r is distance per time, this is a 2nd order differential equation of the form $x''+c_1x'+c_2=0.$.

Experimentally it is found that the velocity component of rotational flow is proportional to r^2, therefore:

$$V_t \propto r^2.$$

This means that the established feed velocity controls the vortex rate inside the cyclone, and the velocity at an arbitrary radius is therefore:

$$Ur=U_{in}\frac{r}{R_{in}}.$$

Subsequently, given a value for V_t, possibly based upon the injection angle, and a cutoff radius, a characteristic particle filtering radius can be estimated, above which particles will be removed from the gas stream.

Alternative Models

The above equations are limited in many regards. For example, the geometry of the separator is not considered, the particles are assumed to achieve a steady state and the effect of the vortex inversion at the base of the cyclone is also ignored, all behaviours which are unlikely to be achieved in a cyclone at real operating conditions.

More complete models exist, as many authors have studied the behaviour of cyclone separators. Numerical modelling using computational fluid dynamics has also been used extensively in the study of cyclonic behaviour. A major limitation of any fluid mechanics model for cyclone separators is the inability to predict the agglomeration of fine particles with larger particles, which has a great impact on cyclone collection efficiency.

Electrostatic Precipitator

Electrodes inside electrostatic precipitator

An electrostatic precipitator (ESP) is a filtration device that removes fine particles, like dust and smoke, from a flowing gas using the force of an induced electrostatic charge minimally impeding the flow of gases through the unit.

Collection electrode of electrostatic precipitator in waste incineration plant

In contrast to wet scrubbers which apply energy directly to the flowing fluid medium, an ESP applies energy only to the particulate matter being collected and therefore is very efficient in its consumption of energy (in the form of electricity).

Invention of the Electrostatic Precipitator

The first use of corona discharge to remove particles from an aerosol was by Hohlfeld in 1824. However, it was not commercialized until almost a century later.

In 1907 Frederick Gardner Cottrell, a professor of chemistry at the University of California, Berkeley, applied for a patent on a device for charging particles and then collecting them through electrostatic attraction—the first electrostatic precipitator. Cottrell first applied the device to the collection of sulphuric acid mist and lead oxide fumes emitted from various acid-making and smelting activities. Wine-producing vineyards in northern California were being adversely affected by the lead emissions.

At the time of Cottrell's invention, the theoretical basis for operation was not understood. The operational theory was developed later in Germany, with the work of Walter Deutsch and the formation of the Lurgi company.

Cottrell used proceeds from his invention to fund scientific research through the creation of a foundation called Research Corporation in 1912, to which he assigned the patents. The intent of the organization was to bring inventions made by educators (such as Cottrell) into the commercial world for the benefit of society at large. The operation of Research Corporation is funded by royalties paid by commercial firms after commercialization occurs. Research Corporation has provided vital funding to many scientific projects: Goddard's rocketry experiments, Lawrence's cyclotron, production methods for vitamins A and B_1, among many others.

By a decision of the US Supreme Court, the Corporation had to be split into several entities. The Research Corporation was separated from two commercial firms making the hardware: Research-Cottrell Inc. (operating east of the Mississippi River) and Western Precipitation (operating in the western states). The Research Corporation continues to be active to this day, and the two companies formed to commercialize the invention for industrial and utility applications are still in business as well.

Electrophoresis is the term used for migration of gas-suspended charged particles in a direct-current electrostatic field. Traditional CRT television sets tend to accumulate dust on the screen because of this phenomenon (a CRT is a direct-current machine operating at about 35 kilovolts).

Plate Precipitator

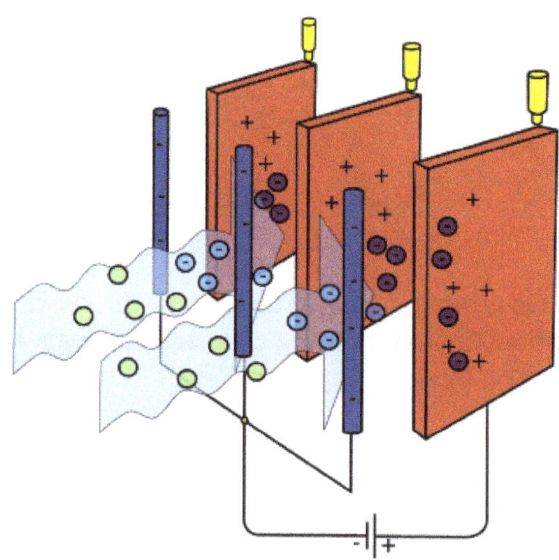

Conceptual diagram of an electrostatic precipitator

The most basic precipitator contains a row of thin vertical wires, and followed by a stack of large flat metal plates oriented vertically, with the plates typically spaced about 1 cm to 18 cm apart, depending on the application. The air stream flows horizontally through the spaces between the wires, and then passes through the stack of plates.

A negative voltage of several thousand volts is applied between wire and plate. If the applied voltage is high enough, an electric corona discharge ionizes the air around the electrodes, which then ionizes the particles in the air stream.

The ionized particles, due to the electrostatic force, are diverted towards the grounded plates. Particles build up on the collection plates and are removed from the air stream.

A two-stage design (separate charging section ahead of collecting section) has the benefit of minimizing ozone production, which would adversely affect health of personnel working in enclosed spaces. For shipboard engine rooms where gearboxes generate an oil mist, two-stage ESP's are used to clean the air, improving the operating environment and preventing buildup of flammable oil fog accumulations. Collected oil is returned to the gear lubricating system.

Collection Efficiency (*R*)

Precipitator performance is very sensitive to two particulate properties: 1) Electrical resistivity; and 2) Particle size distribution. These properties can be measured economically and accurately in the laboratory, using standard tests. Resistivity can be determined as a function of temperature in accordance with IEEE Standard 548. This test is conducted in an air environment containing a specified moisture concentration. The test is run as a function of ascending or descending temperature, or both. Data is acquired using an average ash layer[further explanation needed] electric field of 4 kV/cm. Since relatively low applied voltage is used and no sulfuric acid vapor is present in the test environment, the values obtained indicate the maximum ash resistivity.

In an ESP, where particle charging and discharging are key functions, resistivity is an important factor that significantly affects collection efficiency. While resistivity is an important phenomenon in the inter-electrode region where most particle charging takes place, it has a particularly important effect on the dust layer at the collection electrode where discharging occurs. Particles that exhibit high resistivity are difficult to charge. But once charged, they do not readily give up their acquired charge on arrival at the collection electrode. On the other hand, particles with low resistivity easily become charged and readily release their charge to the grounded collection plate. Both extremes in resistivity impede the efficient functioning of ESPs. ESPs work best under normal resistivity conditions.

Resistivity, which is a characteristic of particles in an electric field, is a measure of a particle's resistance to transferring charge (both accepting and giving up charges). Resistivity is a function of a particle's chemical composition as well as flue gas operating conditions such as temperature and moisture. Particles can have high, moderate (normal), or low resistivity.

Bulk resistivity is defined using a more general version of Ohm's Law, as given in Equation (1) below:

$$\vec{E} = \rho \vec{j} \tag{1}$$

Where:

E is the Electric field strength (V/cm);

j is the Current density (A/cm^2); and

ρ is the Resistivity (Ohm-cm)

A better way of displaying this would be to solve for resistivity as a function of applied voltage and current, as given in Equation (2) below:

$$\rho = \frac{AV}{Il} \tag{2}$$

Where:

ρ = Resistivity (Ohm-cm)

V = The applied DC potential, (Volts);

I = The measured current, (Amperes);

l = The ash layer thickness, (cm); and

A = The current measuring electrode face area, (cm^2).

Resistivity is the electrical resistance of a dust sample 1.0 cm^2 in cross-sectional area, 1.0 cm thick, and is recorded in units of ohm-cm. A method for measuring resistivity will be described in this article. The table below, gives value ranges for low, normal, and high resistivity.

Resistivity	**Range of Measurement**
Low	between 10^4 and 10^7 ohm-cm
Normal	between 10^7 and 2×10^{10} ohm-cm
High	above 2×10^{10} ohm-cm

Dust Layer Resistivity

Resistivity affects electrical conditions in the dust layer by a potential electric field (voltage drop) being formed across the layer as negatively charged particles arrive at its surface and leak their electrical charges to the collection plate. At the metal surface of the electrically grounded collection plate, the voltage is zero, whereas at the outer surface of the dust layer, where new particles and ions are arriving, the electrostatic voltage caused by the gas ions can be quite high. The strength of this electric field depends on the resistivity and thickness of the dust layer.

In high-resistivity dust layers, the dust is not sufficiently conductive, so electrical charges have difficulty moving through the dust layer. Consequently, electrical charges accumulate on and beneath the dust layer surface, creating a strong electric field.

Voltages can be greater than 10,000 volts. Dust particles with high resistivities are held too strongly to the plate, making them difficult to remove and causing rapping problems.

In low resistivity dust layers, the corona current is readily passed to the grounded collection electrode. Therefore, a relatively weak electric field, of several thousand volts, is maintained across the dust layer. Collected dust particles with low resistivity do not adhere strongly enough to the collection plate. They are easily dislodged and become re-entrained in the gas stream.

The electrical conductivity of a bulk layer of particles depends on both surface and volume factors. Volume conduction, or the motions of electrical charges through the interiors of particles, depends mainly on the composition and temperature of the particles. In the higher temperature regions, above 500 °F (260 °C), volume conduction controls the conduction mechanism. Volume conduction also involves ancillary factors, such as compression of the particle layer, particle size and shape, and surface properties.

Volume conduction is represented in the figures as a straight-line at temperatures above 500 °F (260 °C). At temperatures below about 450 °F (230 °C), electrical charges begin to flow across surface moisture and chemical films adsorbed onto the particles. Surface conduction begins to lower the resistivity values and bend the curve downward at temperatures below 500 °F (260 °C).

These films usually differ both physically and chemically from the interiors of the particles owing to adsorption phenomena. Theoretical calculations indicate that moisture films only a few molecules thick are adequate to provide the desired surface conductivity. Surface conduction on particles is closely related to surface-leakage currents occurring on electrical insulators, which have been extensively studied. An interesting practical application of surface-leakage is the determination of dew point by measurement of the current between adjacent electrodes mounted on a glass surface. A sharp rise in current signals the formation of a moisture film on the glass. This method has been used effectively for determining the marked rise in dew point, which occurs when small amounts of sulfuric acid vapor are added to an atmosphere (commercial Dewpoint Meters are available on the market).

The following discussion of normal, high, and low resistivity applies to ESPs operated in a dry state; resistivity is not a problem in the operation of wet ESPs because of the moisture concentration in the ESP. The relationship between moisture content and resistivity is explained later in this work.

Normal Resistivity

As stated above, ESPs work best under normal resistivity conditions. Particles with normal resistivity do not rapidly lose their charge on arrival at the collection electrode. These particles slowly leak their charge to grounded plates and are retained on the collection plates by intermolecular adhesive and cohesive forces. This allows a particulate layer to be built up and then dislodged from the plates by rapping. Within the range of normal dust resistivity (between 10^7 and 2×10^{10} ohm-cm), fly ash is collected more easily than dust having either low or high resistivity.

High Resistivity

If the voltage drop across the dust layer becomes too high, several adverse effects can occur. First, the high voltage drop reduces the voltage difference between the discharge electrode and collection electrode, and thereby reduces the electrostatic field strength used to drive the gas ion-charged particles over to the collected dust layer. As the dust layer builds up, and the electrical charges accumulate on the surface of the dust layer, the voltage difference between the discharge and collection electrodes decreases. The migration velocities of small particles are especially affected by the reduced electric field strength.

Another problem that occurs with high resistivity dust layers is called back corona. This occurs when the potential drop across the dust layer is so great that corona discharges begin to appear in the gas that is trapped within the dust layer. The dust layer breaks down electrically, producing small holes or craters from which back corona discharges occur. Positive gas ions are generated within the dust layer and are accelerated toward the "negatively charged" discharge electrode. The positive ions reduce some of the negative charges on the dust layer and neutralize some of the negative ions on the "charged particles" heading toward the collection electrode. Disruptions of the normal corona process greatly reduce the ESP's collection efficiency, which in severe cases, may fall below 50% . When back corona is present, the dust particles build up on the electrodes forming a layer of insulation. Often this can not be repaired without bringing the unit offline.

The third, and generally most common problem with high resistivity dust is increased electrical sparking. When the sparking rate exceeds the "set spark rate limit," the automatic controllers limit the operating voltage of the field. This causes reduced particle charging and reduced migration velocities toward the collection electrode. High resistivity can generally be reduced by doing the following:

- Adjusting the temperature;
- Increasing moisture content;
- Adding conditioning agents to the gas stream;
- Increasing the collection surface area; and
- Using hot-side precipitators (occasionally and with foreknowledge of sodium depletion).

Thin dust layers and high-resistivity dust especially favor the formation of back corona craters. Severe back corona has been observed with dust layers as thin as 0.1 mm, but a dust layer just over one particle thick can reduce the sparking voltage by 50%. The most marked effects of back corona on the current-voltage characteristics are:

1. Reduction of the spark over voltage by as much as 50% or more;
2. Current jumps or discontinuities caused by the formation of stable back-corona craters; and
3. Large increase in maximum corona current, which just below spark over corona gap may be several times the normal current.

The Figure below and to the left shows the variation in resistivity with changing gas temperature for six different industrial dusts along with three coal-fired fly ashes. The Figure on the right illustrates resistivity values measured for various chemical compounds that were prepared in the laboratory.

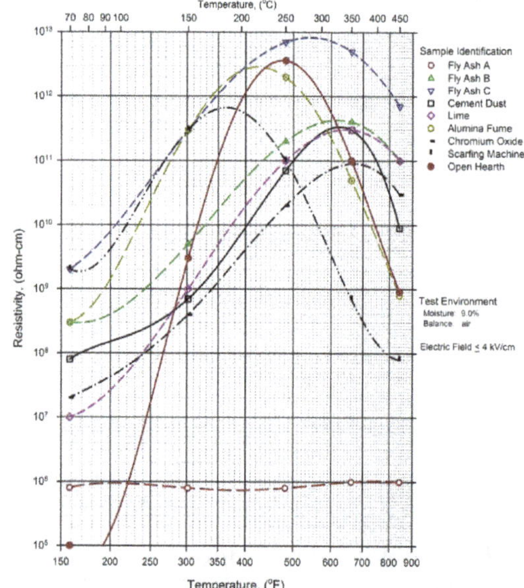

Resistivity Values of Representative Dusts and Fumes From Industrial Plants

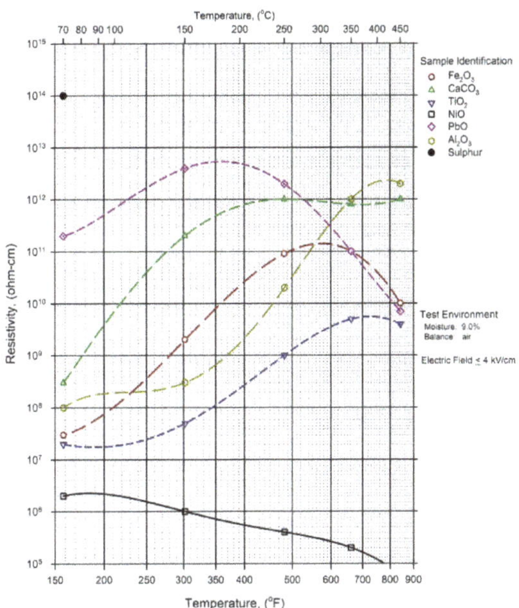

Resistivity Values of Various Chemicals and Reagents as a Function of Temperature

Results for Fly Ash A (in the figure to the left) were acquired in the ascending temperature mode. These data are typical for a moderate to high combustibles content ash. Data for Fly Ash B are from the same sample, acquired during the descending temperature mode.

The differences between the ascending and descending temperature modes are due to the presence of unburned combustibles in the sample. Between the two test modes, the samples are equilibrated in dry air for 14 hours (overnight) at 850 °F (450 °C). This overnight annealing process typically removes between 60% and 90% of any unburned combustibles present in the samples. Exactly how carbon works as a charge carrier is not fully understood, but it is known to significantly reduce the resistivity of a dust.

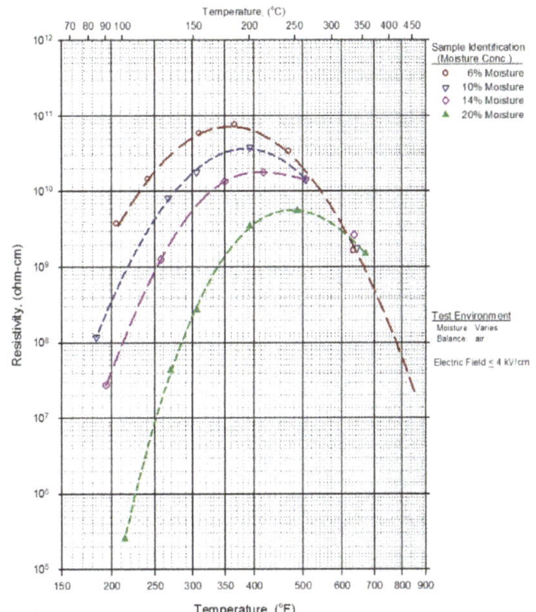

Resistivity Measured as a Function of Temperature in Varying Moisture Concentrations (Humidity)

Carbon can act, at first, like a high resistivity dust in the precipitator. Higher voltages can be required in order for corona generation to begin. These higher voltages can be problematic for the TR-Set controls. The problem lies in onset of corona causing large amounts of current to surge through the (low resistivity) dust layer. The controls sense this surge as a spark. As precipitators are operated in spark-limiting mode, power is terminated and the corona generation cycle re-initiates. Thus, lower power (current) readings are noted with relatively high voltage readings.

The same thing is believed to occur in laboratory measurements. Parallel plate geometry is used in laboratory measurements without corona generation. A stainless steel cup holds the sample. Another stainless steel electrode weight sits on top of the sample (direct contact with the dust layer). As voltage is increased from small amounts (e.g. 20 V), no current is measured. Then, a threshold voltage level is reached. At this level, current surges through the sample... so much so that the voltage supply unit can trip off. After removal of the unburned combustibles during the above-mentioned annealing procedure, the descending temperature mode curve shows the typical inverted "V" shape one might expect.

Low Resistivity

Particles that have low resistivity are difficult to collect because they are easily charged (very conductive) and rapidly lose their charge on arrival at the collection electrode. The particles take on the charge of the collection electrode, bounce off the plates, and become re-entrained in the gas stream. Thus, attractive and repulsive electrical forces that are normally at work at normal and higher resistivities are lacking, and the binding forces to the plate are considerably lessened. Examples of low-resistivity dusts are unburned carbon in fly ash and carbon black.

If these conductive particles are coarse, they can be removed upstream of the precipitator by using a device such as a cyclone mechanical collector.

The addition of liquid ammonia (NH_3) into the gas stream as a conditioning agent has found wide use in recent years. It is theorized that ammonia reacts with H_2SO_4 contained in the flue gas to form an ammonium sulfate compound that increases the cohesivity of the dust. This additional cohesivity makes up for the loss of electrical attraction forces.

The table below summarizes the characteristics associated with low, normal and high resistivity dusts.

The moisture content of the flue gas stream also affects particle resistivity. Increasing the moisture content of the gas stream by spraying water or injecting steam into the duct work preceding the ESP lowers the resistivity. In both temperature adjustment and moisture conditioning, one must maintain gas conditions above the dew point to prevent corrosion problems in the ESP or downstream equipment. The figure to the right shows the effect of temperature and moisture on the resistivity of a cement dust. As the percentage of moisture in the gas stream increases from 6 to 20%, the resistivity of the dust dramatically decreases. Also, raising or lowering the temperature can decrease cement dust resistivity for all the moisture percentages represented.

The presence of SO_3 in the gas stream has been shown to favor the electrostatic precipitation process when problems with high resistivity occur. Most of the sulfur content in the coal burned for combustion sources converts to SO_2. However, approximately 1% of the sulfur converts to SO_3. The amount of SO_3 in the flue gas normally increases with increasing sulfur content of the coal. The resistivity of the particles decreases as the sulfur content of the coal increases.

Resistivity	Range of Measurement	Precipitator Characteristics
Low	between 10^4 and 10^7 ohm-cm	1. Normal operating voltage and current levels unless dust layer is thick enough to reduce plate clearances and cause higher current levels. 2. Reduced electrical force component retaining collected dust, vulnerable to high reentrainment losses. 3. Negligible voltage drop across dust layer. 4. Reduced collection performance due to (2)
Normal	between 10^7 and 2 x 10^{10} ohm-cm	1. Normal operating voltage and current levels. 2. Negligible voltage drop across dust layer. 3. Sufficient electrical force component retaining collected dust. 4. High collection performance due to (1), (2) and (3)
Marginal to High	between 2 x 10^{10} and 10^{12} ohm-cm	1. Reduced operating voltage and current levels with high spark rates. 2. Significant voltage loss across dust layer. 3. Moderate electrical force component retaining collected dust. 4. Reduced collection performance due to (1) and (2)
High	above 10^{12} ohm-cm	1. Reduced operating voltage levels; high operating current levels if power supply controller is not operating properly. 2. Very significant voltage loss across dust layer. 3. High electrical force component retaining collected dust. 4. Seriously reduced collection performance due to (1), (2) and probably back corona.

Other conditioning agents, such as sulfuric acid, ammonia, sodium chloride, and soda ash (sometimes as raw trona), have also been used to reduce particle resistivity. Therefore, the chemical composition of the flue gas stream is important with regard to the resistivity of the particles to be collected in the ESP. The table below lists various conditioning agents and their mechanisms of operation.

Conditioning Agent	Mechanism(s) of Action
Sulfur Trioxide and/or Sulfuric Acid	1. Condensation and adsorption on fly ash surfaces. 2. may also increase cohesiveness of fly ash. 3. Reduces resistivity
Ammonia	Mechanism is not clear, various ones proposed; 1. Modifies resistivity. 2. Increases ash cohesiveness. 3. Enhances space charge effect.

Ammonium Sulfate	Little is known about the mechanism; claims are made for the following: 1. Modifies resistivity (depends upon injection temperature). 2. Increases ash cohesiveness. 3. Enhances space charge effect. 4. Experimental data lacking to substantiate which of these is predominant.
Triethylamine	Particle agglomeration claimed; no supporting data.
Sodium Compounds	1. Natural conditioner if added with coal. 2. Resistivity modifier if injected into gas stream.
Compounds of Transition Metals	Postulated that they catalyze oxidation of SO_2 to SO_3; no definitive tests with fly ash to verify this postulation.
Potassium Sulfate and Sodium Chloride	In cement and lime kiln ESPs: 1. Resistivity modifiers in the gas stream. 2. NaCl - natural conditioner when mixed with coal.

If injection of ammonium sulfate occurs at a temperature greater than about 600 °F (320 °C), dissociation into ammonia and sulfur trioxide results. Depending on the ash, SO_2 may preferentially interact with fly ash as SO_3 conditioning. The remainder recombines with ammonia to add to the space charge as well as increase cohesiveness of the ash.

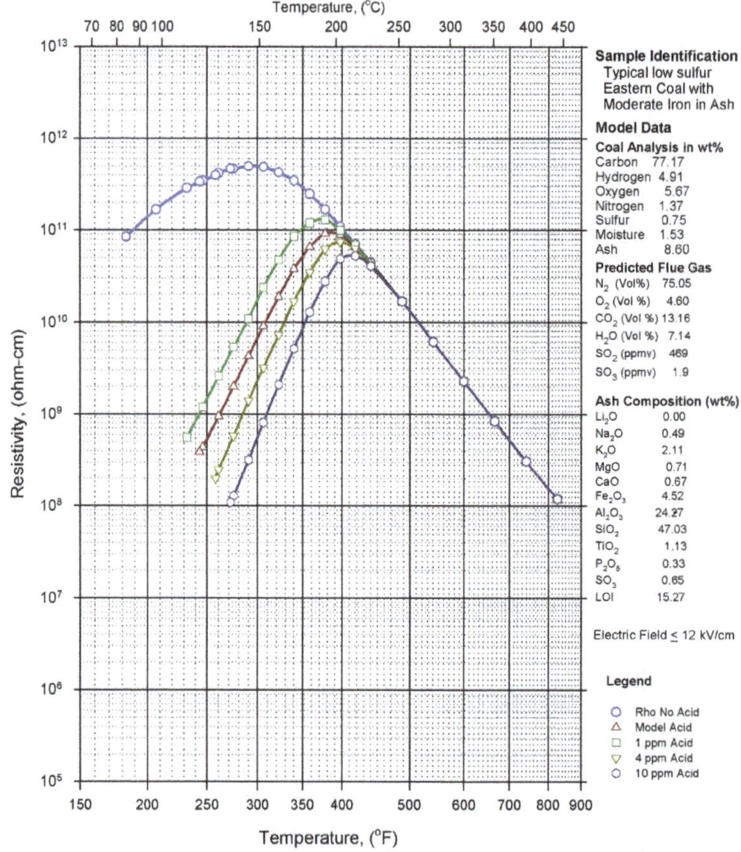

Resistivity Modeled As A Function of Environmental Conditions - Especially Sulfuric Acid Vapor

More recently, it has been recognized that a major reason for loss of efficiency of the electrostatic precipitator is due to particle buildup on the charging wires in addition to the collection plates (Davidson and McKinney, 1998). This is easily remedied by making sure that the wires themselves are cleaned at the same time that the collecting plates are cleaned.

Sulfuric acid vapor (SO_3) enhances the effects of water vapor on surface conduction. It is physically adsorbed within the layer of moisture on the particle surfaces. The effects of relatively small amounts of acid vapor can be seen in the figure below and to the right.

The inherent resistivity of the sample at 300 °F (150 °C) is 5×10^{12} ohm-cm. An equilibrium concentration of just 1.9 ppm sulfuric acid vapor lowers that value to about 7×10^9 ohm-cm.

Modern Industrial Electrostatic Precipitators

A smokestack at coal-fired Hazelwood Power Station in Victoria, Australia emits brown smoke when its ESP is shut down

ESPs continue to be excellent devices for control of many industrial particulate emissions, including smoke from electricity-generating utilities (coal and oil fired), salt cake collection from black liquor boilers in pulp mills, and catalyst collection from fluidized bed catalytic cracker units in oil refineries to name a few. These devices treat gas volumes from several hundred thousand ACFM to 2.5 million ACFM (1,180 m^3/s) in the largest coal-fired boiler applications. For a coal-fired boiler the collection is usually performed downstream of the air preheater at about 160 °C (320 °F) which provides optimal resistivity of the coal-ash particles. For some difficult applications with low-sulfur fuel hot-end units have been built operating above 370 °C (698 °F).

The original parallel plate–weighted wire design has evolved as more efficient (and robust) discharge electrode designs were developed, today focusing on rigid (pipe-frame) discharge electrodes to which many sharpened spikes are attached (barbed wire), maximizing corona production. Transformer-rectifier systems apply voltages of 50–100 kV at relatively high current densities. Modern controls, such as an automatic voltage control, minimize electric sparking and prevent arcing (sparks are quenched within 1/2 cycle of the TR set), avoiding damage to the components. Automatic plate-rapping systems and hopper-evacuation systems remove the collected particulate matter while on line, theoretically allowing ESPs to stay in continuous operation for years at a time.

Electrostatic Air Sampling

Electrostatic precipitators can be used to sample airborne particles or aerosol for analysis purposes. Precipitator designs with a liquid counterelectrode can be used to sample biological particles, e.g. viruses, directly into a small liquid volume to reduce unnecessary sample dilution.

Wet Electrostatic Precipitator

A wet electrostatic precipitator (WESP or wet ESP) operates with water vapor saturated air streams (100% relative humidity). WESPs are commonly used to remove liquid droplets such as sulfuric acid mist from industrial process gas streams. The WESP is also commonly used where the gases are high in moisture content, contain combustible particulate, or have particles that are sticky in nature.

The preferred and most modern type of WESP is a downflow tubular design. This design allows the collected moisture and particulate to form a moving slurry that helps to keep the collection surfaces clean. Plate style and upflow design WESPs are very unreliable and should not be used in applications where particulate is sticky in nature.

Consumer-oriented Electrostatic Air Cleaners

A portable electrostatic air cleaner marketed to consumers

Plate precipitators are commonly marketed to the public as air purifier devices or as a permanent replacement for furnace filters, but all have the undesirable attribute of being somewhat messy to clean. A negative side-effect of electrostatic precipitation devices is the potential production of toxic ozone and NO_x. However, electrostatic precipitators offer benefits over other air purifications technologies, such as HEPA filtration, which require expensive filters and can become "production sinks" for many harmful forms of bacteria.

With electrostatic precipitators, if the collection plates are allowed to accumulate large amounts of particulate matter, the particles can sometimes bond so tightly to the metal plates that vigorous washing and scrubbing may be required to completely clean the collection plates. The close spacing of the plates can make thorough cleaning difficult, and the stack of plates often cannot be easily disassembled for cleaning. One solution, suggested by several manufacturers, is to wash the collector plates in a dishwasher.

Portable electrostatic air cleaner with cover removed, showing collector plates

Some consumer precipitation filters are sold with special soak-off cleaners, where the entire plate array is removed from the precipitator and soaked in a large container overnight, to help loosen the tightly bonded particulates.

A study by the Canada Mortgage and Housing Corporation testing a variety of forced-air furnace filters found that ESP filters provided the best, and most cost-effective means of cleaning air using a forced-air system.

The first portable electrostatic air filter systems for homes was marketed in 1954 by Raytheon.

Dust Collector

A dust collector is a system used to enhance the quality of air released from industrial and commercial processes by collecting dust and other impurities from air or gas. Designed to handle high-volume dust loads, a dust collector system consists of a blower, dust filter, a filter-cleaning system, and a dust receptacle or dust removal system. It is distinguished from air cleaners, which use disposable filters to remove dust.

Baghouse Dust Collector for Asphalt Plants

Two rooftop dust collectors in Pristina, Kosovo

History

Wilhelm Beth

„Beth"-Filter „KS" (1910)

The father of the dust collector was Wilhelm Beth from Lübeck.

Uses

Dust collectors are used in many processes to either recover valuable granular solid or powder from process streams, or to remove granular solid pollutants from exhaust gases prior to venting to the atmosphere. Dust collection is an online process for collecting any process-generated dust from the source point on a continuous basis. Dust collectors may be of single unit construction, or a collection of devices used to separate particulate matter from the process air. They are often used as an air pollution control device to maintain or improve air quality.

Mist collectors remove particulate matter in the form of fine liquid droplets from the air. They are often used for the collection of metal working fluids, and coolant or oil mists. Mist collectors are often used to improve or maintain the quality of air in the workplace environment.

Fume and smoke collectors are used to remove sub-micrometer-size particulates from the air. They effectively reduce or eliminate particulate matter and gas streams from many industrial processes such as welding, rubber and plastic processing, high speed machining with coolants, tempering, and quenching.

Types of Dust Collectors

Five main types of industrial dust collectors are:

- Inertial separators
- Fabric filters

- Wet scrubbers
- Unit collectors
- Electrostatic precipitators

Inertial Separators

Inertial separators separate dust from gas streams using a combination of forces, such as centrifugal, gravitational, and inertial. These forces move the dust to an area where the forces exerted by the gas stream are minimal. The separated dust is moved by gravity into a hopper, where it is temporarily stored.

The three primary types of inertial separators are:

- Settling chambers
- Baffle chambers
- Centrifugal collectors

Neither settling chambers nor baffle chambers are commonly used in the minerals processing industry. However, their principles of operation are often incorporated into the design of more efficient dust collectors.

Settling Chamber

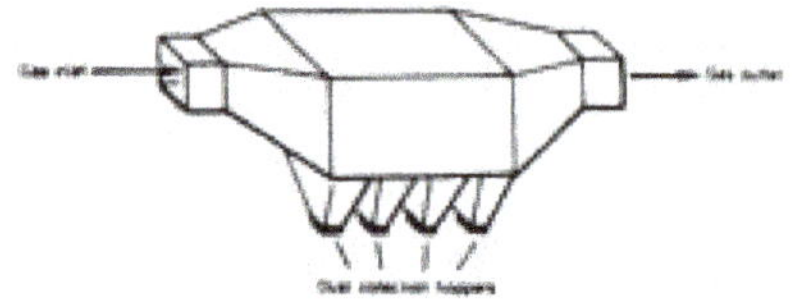

Settling Chamber

A settling chamber consists of a large box installed in the ductwork. The increase of cross section area at the chamber reduces the speed of the dust-filled airstream and heavier particles settle out.

Settling chambers are simple in design and can be manufactured from almost any material. However, they are seldom used as primary dust collectors because of their large space requirements and low efficiency. A practical use is as precleaners for more efficient collectors.

Baffle Chamber

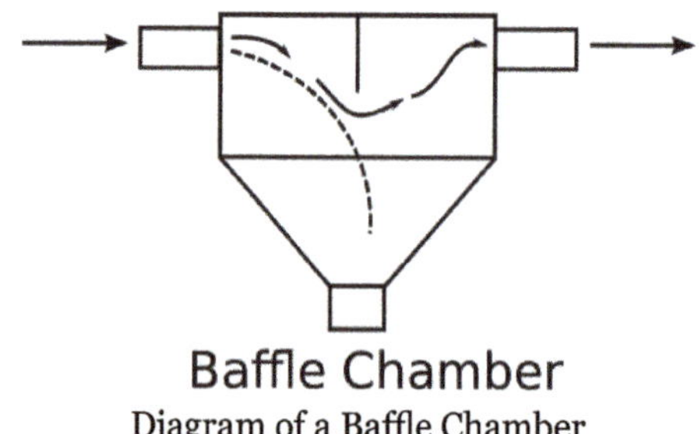

Diagram of a Baffle Chamber

Baffle chambers use a fixed baffle plate that causes the conveying gas stream to make a sudden change of direction. Large-diameter particles do not follow the gas stream but continue into a dead air space and settle. Baffle chambers are used as precleaners

Centrifugal Collectors

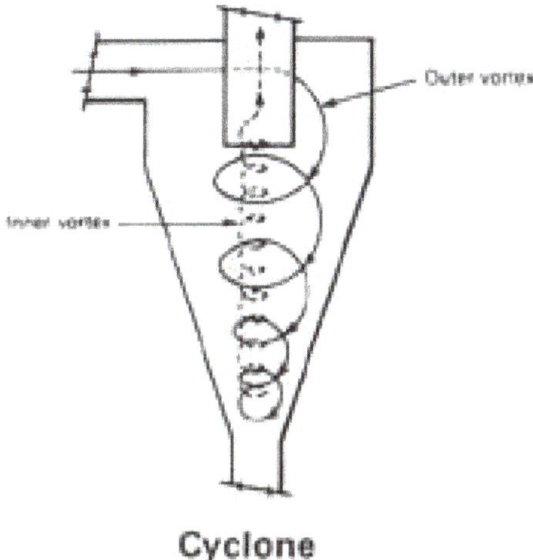

Cyclone

Centrifugal collectors use cyclonic action to separate dust particles from the gas stream. In a typical cyclone, the dust gas stream enters at an angle and is spun rapidly. The centrifugal force created by the circular flow throws the dust particles toward the wall of the cyclone. After striking the wall, these particles fall into a hopper located underneath.

The most common types of centrifugal, or inertial, collectors in use today are:-

Single-cyclone Separators

They create a dual vortex to separate coarse from fine dust. The main vortex spirals downward and carries most of the coarser dust particles. The inner vortex, created near the bottom of the cyclone, spirals upward and carries finer dust particles.

Multiple-cyclone Separators

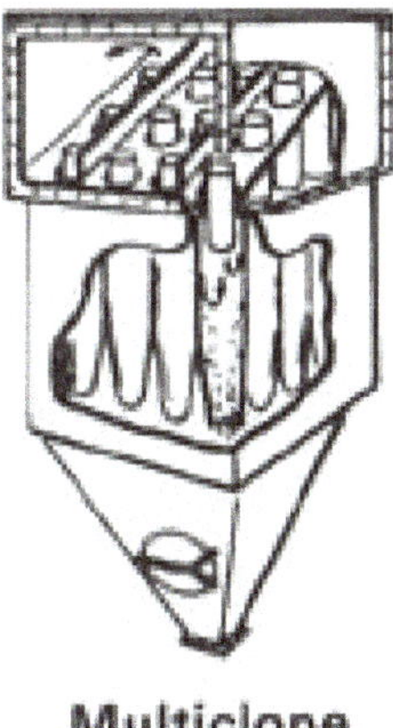

Multiclone

Multiple-cyclone separators consist of a number of small-diameter cyclones, operating in parallel and having a common gas inlet and outlet, as shown in the figure, and operate on the same principle as single cyclone separators—creating an outer downward vortex and an ascending inner vortex.

Multiple-cyclone separators remove more dust than single cyclone separators because the individual cyclones have a greater length and smaller diameter. The longer length provides longer residence time while the smaller diameter creates greater centrifugal force. These two factors result in better separation of dust particulates. The pressure drop of multiple-cyclone separators collectors is higher than that of single-cyclone separators, requiring more energy to clean the same amount of air. A single-chamber cyclone separator of the same volume is more economical, but doesn't remove as much dust.

Cyclone separators are found in all types of power and industrial applications, including pulp and paper plants, cement plants, steel mills, petroleum coke plants, metallurgical plants, saw mills and other kinds of facilities that process dust.

Secondary-air-flow Separators

This type of cyclone uses a secondary air flow, injected into the cyclone to accomplish several things. The secondary air flow increases the speed of the cyclonic action making the separator more efficient; it intercepts the particulate before it reaches the interior walls of the unit; and it forces the separated particulate toward the collection area. The secondary air flow protects the separator from particulate abrasion and allows the separator to be installed horizontally because gravity is not depended upon to move the separated particulate downward.

Fabric Filters

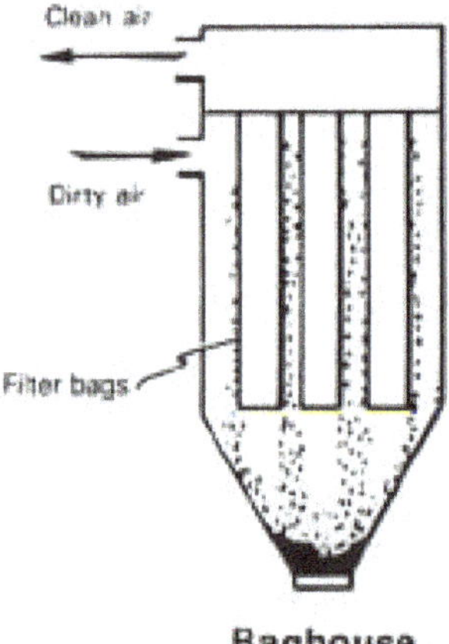

Baghouse

Commonly known as baghouses, fabric collectors use filtration to separate dust particulates from dusty gases. They are one of the most efficient and cost effective types of dust collectors available and can achieve a collection efficiency of more than 99% for very fine particulates.

Dust-laden gases enter the baghouse and pass through fabric bags that act as filters. The bags can be of woven or felted cotton, synthetic, or glass-fiber material in either a tube or envelope shape.

Pre-coating

To ensure the filter bags have a long usage life they are commonly coated with a filter enhancer (pre-coat). The use of chemically inert limestone (calcium carbonate) is most common as it maximises efficiency of dust collection (including fly ash) via formation of what is called a dustcake or coating on the surface of the filter media. This not only traps fine particulates but also provides protection for the bag itself from moisture, and oily or sticky particulates which can bind the filter media. Without a pre-coat the filter bag allows fine particulates to bleed through the bag filter system, especially during start-up, as the bag can only do part of the filtration leaving the finer parts to the filter enhancer dustcake.

Parts

Fabric filters generally have the following parts:

1. Clean plenum
2. Dusty plenum
3. Bag, cage, venturi assembly
4. Tubeplate
5. RAV/SCREW
6. Compressed air header
7. Blow pipe
8. Housing and hopper

Types of Bag Cleaning

Baghouses are characterized by their cleaning method.

Shaking

A rod connecting to the bag is powered by a motor. This provides motion to remove caked-on particles. The speed and motion of the shaking depends on the design of the bag and composition of the particulate matter. Generally shaking is horizontal. The top of the bag is closed and the bottom is open. When shaken, the dust collected on the inside of the bag is freed. During the cleaning process, no dirty gas flows through a bag while the bag is being cleaned. This redirection of air flow illustrates why baghouses must be compartmentalized.

Reverse Air

Air flow gives the bag structure. Dirty air flows through the bag from the inside, allowing dust to collect on the interior surface. During cleaning, gas flow is restricted from a specific compartment. Without the flowing air, the bags relax. The cylindrical bag contains rings that prevent it from completely collapsing under the pressure of the air. A fan blows clean air in the reverse direction. The relaxation and reverse air flow cause the dust cake to crumble and release into the hopper. Upon the completion of the cleaning process, dirty air flow continues and the bag regains its shape.

Pulse Jet

This type of baghouse cleaning (also known as pressure-jet cleaning) is the most common. A high pressure blast of air is used to remove dust from the bag. The blast enters the top of the bag tube, temporarily ceasing the flow of dirty air. The shock of air causes a wave of expansion to travel down the fabric. The flexing of the bag shatters and discharges the dust cake. The air burst is about 0.1 second and it takes about 0.5 seconds for the shock wave to travel down the length of the bag. Due to its rapid release, the blast of air does not interfere with contaminated gas flow. Therefore, pulse-jet baghouses can operate continuously and are not usually compartmentalized. The blast of compressed air must be powerful enough to ensure that the shock wave will travel the entire length of the bag and fracture the dust cake.

Sonic

The least common type of cleaning method is sonic. Shaking is achieved by sonic vibration. A sound generator produces a low frequency sound that causes the bags to vibrate. Sonic cleaning is commonly combined with another method of cleaning to ensure thorough cleaning.

Cartridge Collectors

Cartridge collectors use perforated metal cartridges that contain a pleated, nonwoven filtering media, as opposed to woven or felt bags used in baghouses. The pleated design allows for a greater total filtering surface area than in a conventional bag of the same diameter, The greater filtering area results in a reduced air to media ratio, pressure drop, and overall collector size.

Cartridge collectors are available in single use or continuous duty designs. In single-use collectors, the dirty cartridges are changed and collected dirt is removed while the collector is off. In the continuous duty design, the cartridges are cleaned by the conventional pulse-jet cleaning system.

Wet Scrubbers

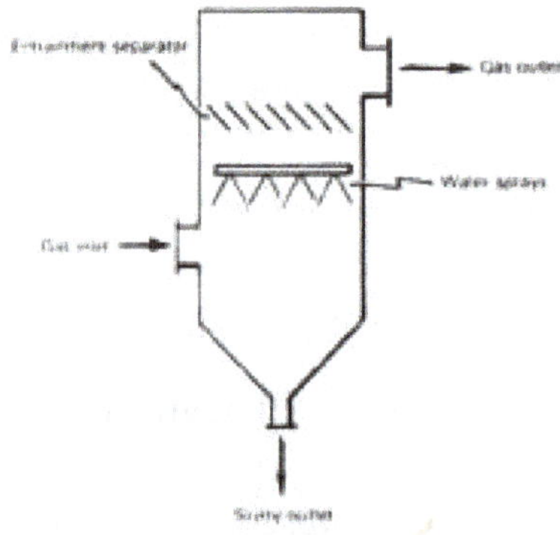

Wet Scrubber

Dust collectors that use liquid are known as wet scrubbers. In these systems, the scrubbing liquid (usually water) comes into contact with a gas stream containing dust particles. Greater contact of the gas and liquid streams yields higher dust removal efficiency.

There is a large variety of wet scrubbers; however, all have one of three basic configurations:

1. Gas-humidification - The gas-humidification process agglomerates fine particles, increasing the bulk, making collection easier.

2. Gas-liquid contact - This is one of the most important factors affecting collection efficiency. The particle and droplet come into contact by four primary mechanisms:

a) Inertial impaction - When water droplets placed in the path of a dust-laden gas stream, the stream separates and flows around them. Due to inertia, the larger dust particles will continue on in a straight path, hit the droplets, and become encapsulated.

b) Interception - Finer particles moving within a gas stream do not hit droplets directly but brush against and adhere to them.

c) Diffusion - When liquid droplets are scattered among dust particles, the particles are deposited on the droplet surfaces by Brownian movement, or diffusion. This is the principal mechanism in the collection of submicrometre dust particles.

d) Condensation nucleation - If a gas passing through a scrubber is cooled below the dew-point, condensation of moisture occurs on the dust particles. This increase in particle size makes collection easier.

3. Gas-liquid separation - Regardless of the contact mechanism used, as much liquid and dust as possible must be removed. Once contact is made, dust particulates and water droplets combine to form agglomerates. As the agglomerates grow larger, they settle into a collector.

The "cleaned" gases are normally passed through a mist eliminator (demister pads) to remove water droplets from the gas stream. The dirty water from the scrubber system is either cleaned and discharged or recycled to the scrubber. Dust is removed from the scrubber in a clarification unit or a drag chain tank. In both systems solid material settles on the bottom of the tank. A drag chain system removes the sludge and deposits in into a dumpster or stockpile.

Types of Scrubbers

Spray-tower scrubber wet scrubbers may be categorized by pressure drop as follows:

- Low-energy scrubbers (0.5 to 2.5 inches water gauge - 124.4 to 621.9 Pa)
- Low- to medium-energy scrubbers (2.5 to 6 inches water gauge - 0.622 to 1.493 kPa)
- Medium- to high-energy scrubbers (6 to 15 inches water gauge - 1.493 to 3.731 kPa)
- High-energy scrubbers (greater than 15 inches water gauge - greater than 3.731 kPa)

Due to the large number of commercial scrubbers available, it is not possible to describe each individual type here. However, the following sections provide examples of typical scrubbers in each category.

Low-energy Scrubbers

In the simple, gravity-spray-tower scrubber, liquid droplets formed by liquid atomized in spray nozzles fall through rising exhaust gases. Dirty water is drained at the bottom.

These scrubbers operated at pressure drops of 1 to 2 in. water gauge (¼ to ½ kPa) and are approximately 70% efficient on 10 μm particles. Their efficiency is poor below 10 μm. However, they are capable of treating relatively high dust concentrations without becoming plugged.

Low- to Medium-energy Scrubbers

Wet cyclones use centrifugal force to spin the dust particles (similar to a cyclone), and throw the particulates upon the collector's wetted walls. Water introduced from the top to wet the cyclone walls carries these particles away. The wetted walls also prevent dust reentrainment.

Pressure drops for these collectors range from 2 to 8 in. water (½ to 2 kPa), and the collection efficiency is good for 5 μm particles and above.

High-energy Scrubbers co-current-flow Scrubber

Packed-bed scrubbers consist of beds of packing elements, such as coke, broken rock, rings, saddles, or other manufactured elements. The packing breaks down the liquid flow into a high-surface-area film so that the dusty gas streams passing through the bed achieve maximum contact with the liquid film and become deposited on the surfaces of the packing elements. These scrubbers have a good collection efficiency for respirable dust.

Three types of packed-bed scrubbers are-

- Cross-flow scrubbers
- Co-current flow scrubbers
- Counter-current flow scrubbers

Efficiency can be greatly increased by minimizing target size, i.e., using 0.003 in. (0.076 mm) diameter stainless steel wire and increasing gas velocity to more than 1,800 ft/min (9.14 m/s).

High-energy Scrubbers

Venturi scrubbers consist of a venturi-shaped inlet and separator. The dust-laden gases venturi scrubber enter through the venturi and are accelerated to speeds between 12,000 and 36,000 ft/min (60.97-182.83 m/s). These high-gas velocities immediately atomize the coarse water spray, which is injected radially into the venturi throat, into fine droplets. High energy and extreme turbulence promote collision between water droplets and dust particulates in the throat. The agglomeration process between particle and droplet continues in the diverging section of the venturi. The large agglomerates formed in the venturi are then removed by an inertial separator.

Venturi scrubbers achieve very high collection efficiencies for respirable dust. Since efficiency of a venturi scrubber depends on pressure drop, some manufacturers supply a variable-throat venturi to maintain pressure drop with varying gas flows.

Electrostatic Precipitators (ESP)

Electrostatic precipitators use electrostatic forces to separate dust particles from exhaust gases. A number of high-voltage, direct-current discharge electrodes are placed between grounded collecting electrodes. The contaminated gases flow through the passage formed by the discharge and collecting electrodes. Electrostatic precipitators operate on the same principle as home "Ionic" air purifiers.

The airborne particles receive a negative charge as they pass through the ionized field between the electrodes. These charged particles are then attracted to a grounded or positively charged electrode and adhere to it.

The collected material on the electrodes is removed by rapping or vibrating the collecting electrodes either continuously or at a predetermined interval. Cleaning a precipitator can usually be done without interrupting the airflow.

The four main components of all electrostatic precipitators are:

- Power supply unit, to provide high-voltage DC power
- Ionizing section, to impart a charge to particulates in the gas stream
- A means of removing the collected particulates
- A housing to enclose the precipitator zone

The following factors affect the efficiency of electrostatic precipitators:

- Larger collection-surface areas and lower gas-flow rates increase efficiency because of the increased time available for electrical activity to treat the dust particles.
- An increase in the dust-particle migration velocity to the collecting electrodes increases efficiency. The migration velocity can be increased by:
    - Decreasing the gas viscosity
    - Increasing the gas temperature
    - Increasing the voltage field

Types of Precipitators

There are two main types of precipitators:

- High-voltage, single-stage - Single-stage precipitators combine an ionization and a collection step. They are commonly referred to as Cottrell precipitators.
- Low-voltage, two-stage - Two-stage precipitators use a similar principle; however, the ionizing section is followed by collection plates.

Described below is the high-voltage, single-stage precipitator, which is widely used in minerals processing operations. The low-voltage, two-stage precipitator is generally used for filtration in air-conditioning systems.

Plate Precipitators

The majority of electrostatic precipitators installed are the plate type. Particles are collected on flat, parallel surfaces that are 8 to 12 in. (20 to 30 cm) apart, with a series of discharge electrodes spaced along the centerline of two adjacent plates. The contaminated gases pass through the passage between the plates, and the particles become charged and adhere to the collection plates. Collected particles are usually removed by rapping the plates and deposited in bins or hoppers at the base of the precipitator.

Tubular Precipitators

Tubular precipitators consist of cylindrical collection electrodes with discharge electrodes located on the axis of the cylinder. The contaminated gases flow around the discharge electrode and up through the inside of the cylinders. The charged particles are collected on the grounded walls of the cylinder. The collected dust is removed from the bottom of the cylinder.

Tubular precipitators are often used for mist or fog collection or for adhesive, sticky, radioactive, or extremely toxic materials.

Unit Collectors

Unlike central collectors, unit collectors control contamination at its source. They are small and self-contained, consisting of a fan and some form of dust collector. They are suitable for isolated, portable, or frequently moved dust-producing operations, such as bins and silos or remote belt-conveyor transfer points. Advantages of unit collectors include small space requirements, the return of collected dust to main material flow, and low initial cost. However, their dust-holding and storage capacities, servicing facilities, and maintenance periods have been sacrificed.

A number of designs are available, with capacities ranging from 200 to 2,000 ft^3/min (90 to 900 L/s). There are two main types of unit collectors:

- Fabric collectors, with manual shaking or pulse-jet cleaning - normally used for fine dust
- Cyclone collectors - normally used for coarse dust

Fabric collectors are frequently used in minerals processing operations because they provide high collection efficiency and uninterrupted exhaust airflow between cleaning cycles. Cyclone collectors are used when coarser dust is generated, as in woodworking, metal grinding, or machining.

The following points should be considered when selecting a unit collector:

- Cleaning efficiency must comply with all applicable regulations.
- The unit maintains its rated capacity while accumulating large amounts of dust between cleanings.

- Simple cleaning operations do not increase the surrounding dust concentration.
- Has the ability to operate unattended for extended periods of time (for example, 8 hours).
- Automatic discharge or sufficient dust storage space to hold at least one week's accumulation.
- If renewable filters are used, they should not have to be replaced more than once a month.
- Durable.
- Quiet.

Use of unit collectors may not be appropriate if the dust-producing operations are located in an area where central exhaust systems would be practical. Dust removal and servicing requirements are expensive for many unit collectors and are more likely to be neglected than those for a single, large collector.

Selecting a Dust Collector

Dust collectors vary widely in design, operation, effectiveness, space requirements, construction, and capital, operating, and maintenance costs. Each type has advantages and disadvantages. However, the selection of a dust collector should be based on the following general factors:

- Dust concentration and particle size - For minerals processing operations, the dust concentration can range from 0.1 to 5.0 grains (0.32 g) of dust per cubic feet of air (0.23 to 11.44 grams per standard cubic meter), and the particle size can vary from 0.5 to 100 micrometres (μm) in diameter.
- Degree of dust collection required - The degree of dust collection required depends on its potential as a health hazard or public nuisance, the plant location, the allowable emission rate, the nature of the dust, its salvage value, and so forth. The selection of a collector should be based on the efficiency required and should consider the need for high-efficiency, high-cost equipment, such as electrostatic precipitators; high-efficiency, moderate-cost equipment, such as baghouses or wet scrubbers; or lower cost, primary units, such as dry centrifugal collectors.
- Characteristics of airstream - The characteristics of the airstream can have a significant impact on collector selection. For example, cotton fabric filters cannot be used where air temperatures exceed 180 °F (82 °C). Also, condensation of steam or water vapor can blind bags. Various chemicals can attack fabric or metal and cause corrosion in wet scrubbers.
- Characteristics of dust - Moderate to heavy concentrations of many dusts (such as dust from silica sand or metal ores) can be abrasive to dry centrifugal collectors. Hygroscopic material can blind bag collectors. Sticky material can adhere to collector elements and plug passages. Some particle sizes and shapes may rule out certain types of fabric collectors. The combustible nature of many fine materials rules out the use of electrostatic precipitators.
- Methods of disposal - Methods of dust removal and disposal vary with the material, plant process, volume, and type of collector used. Collectors can unload continuously or in batches. Dry

materials can create secondary dust problems during unloading and disposal that do not occur with wet collectors. Disposal of wet slurry or sludge can be an additional material-handling problem; sewer or water pollution problems can result if wastewater is not treated properly.

Fan and Motor

The fan and motor system supplies mechanical energy to move contaminated air from the dust-producing source to a dust collector.

Types of Fans

There are two main kinds of industrial fans:

- Centrifugal fans
- Axial-flow fans

Centrifugal Fans

Centrifugal fans consist of a wheel or a rotor mounted on a shaft that rotates in a scroll-shaped housing. Air enters at the eye of the rotor, makes a right-angle turn, and is forced through the blades of the rotor by centrifugal force into the scroll-shaped housing. The centrifugal force imparts static pressure to the air. The diverging shape of the scroll also converts a portion of the velocity pressure into static pressure.

There are three main types of centrifugal fans:

- Radial-blade fans - Radial-blade fans are used for heavy dust loads. Their straight, radial blades do not get clogged with material, and they withstand considerable abrasion. These fans have medium tip speeds and medium noise factors.
- Backward-blade fans - Backward-blade fans operate at higher tip speeds and thus are more efficient. Since material may build up on the blades, these fans should be used after a dust collector. Although they are noisier than radial-blade fans, backward-blade fans are commonly used for large-volume dust collection systems because of their higher efficiency.
- Forward-curved-blade fans - These fans have curved blades that are tipped in the direction of rotation. They have low space requirements, low tip speeds, and a low noise factor. They are usually used against low to moderate static pressures.

Axial-flow Fans

Axial-flow fans are used in systems that have low resistance levels. These fans move the air parallel to the fan's axis of rotation. The screw-like action of the propellers moves the air in a straight-through parallel path, causing a helical flow pattern.

The three main kinds of axial fans are-

- Propeller fans - These fans are used to move large quantities of air against very low static pressures. They are usually used for general ventilation or dilution ventilation and are good in developing up to 0.5 in. wg (124.4 Pa).

- Tube-axial fans - Tube-axial fans are similar to propeller fans except they are mounted in a tube or cylinder. Therefore, they are more efficient than propeller fans and can develop up to 3 to 4 in. wg (743.3 to 995 Pa). They are best suited for moving air containing substances such as condensible fumes or pigments.

- Vane-axial fans - Vane-axial fans are similar to tube-axial fans except air-straightening vanes are installed on the suction or discharge side of the rotor. They are easily adapted to multistaging and can develop static pressures as high as 14 to 16 in. wg (3.483 to 3.98 kPa). They are normally used for clean air only.

Fan Selection

When selecting a fan, the following points should be considered:

- Volume required
- Fan static pressure
- Type of material to be handled through the fan (For example, a radial-blade fan should be used with fibrous material or heavy dust loads, and nonsparking construction must be used with explosive or inflammable materials.)
- Type of drive arrangement, such as direct drive or belt drive
- Space requirements
- Noise levels
- Operating temperature (For example, sleeve bearings are suitable to 250 °F/121.1 °C; ball bearings to 550 °F/287.8 °C)
- Sufficient size to handle the required volume and pressure with minimum horsepower
- Need for special coatings or construction when operating in corrosive atmospheres
- Ability of fan to accommodate small changes in total pressure while maintaining the necessary air volume
- Need for an outlet damper to control airflow during cold starts (If necessary, the damper may be interlocked with the fan for a gradual start until steady-state conditions are reached.)

Fan Rating Tables

After the above information is collected, the actual selection of fan size and speed is usually made from a rating table published by the fan manufacturer. This table is known as a multirating table, and it shows the complete range of capacities for a particular size of fan.

Points to note:

- The multirating table shows the range of pressures and speeds possible within the limits of the fan's construction.
- A particular fan may be available in different construction classes (identified as class I through IV) relating to its capabilities and limits.
- For a given pressure, the highest mechanical efficiency is usually found in the middle third of the volume column.
- A fan operating at a given speed can have an infinite number of ratings (pressure and volume) along the length of its characteristic curve. However, when the fan is installed in a dust collection system, the point of rating can only be at the point at which the system resistance curve intersects the fan characteristic curve.
- In a given system, a fan at a fixed speed or at a fixed blade setting can have a single rating only. This rating can be changed only be changing the fan speed, blade setting, or the system resistance.
- For a given system, an increase in exhaust volume will result in increases in static and total pressures. For example, for a 20% increase in exhaust volume in a system with 5 in. pressure loss, the new pressure loss will be $5 \times (1.20)^2 = 7.2$ in.
- For rapid estimates of probable exhaust volumes available for a given motor size, the equation for brake horsepower, as illustrated, can be useful.

Fan installation Typical fan discharge conditions Fan ratings for volume and static pressure, as described in the multirating tables, are based on the tests conducted under ideal conditions. Often, field installation creates airflow problems that reduce the fan's air delivery. The following points should be considered when installing the fan:

- Avoid installation of elbows or bends at the fan discharge, which will lower fan performance by increasing the system's resistance.
- Avoid installing fittings that may cause non-uniform flow, such as an elbow, mitred elbow, or square duct.
- Check that the fan impeller is rotating in the proper direction-clockwise or counterclockwise.
- For belt-driven fans-
    - Check that the motor sheave and fan sheave are aligned properly.
    - Check for proper belt tension.
- Check the passages between inlets, impeller blades, and inside of housing for buildup of dirt, obstructions, or trapped foreign matter.

Electric Motors

Electric motors are used to supply the necessary energy to drive the fan.

Integral-horsepower electric motors are normally three-phase, alternating-current motors. Fractional-horsepower electric motors are normally single-phase, alternating-current motors and are used when less than 1 hp (0.75 kW) is required. Since most dust collection systems require motors with more than 1 hp (0.75 kW), only integral-horsepower motors are discussed here.

The two most common types of integral-horsepower motors used in dust collection systems are-

- Squirrel-cage motors - These motors have a constant speed and are of a nonsynchronous, induction type.
- Wound-rotor motors - These motors are also known as slip-ring motors. They are general-purpose or continuous-rated motors and are chiefly used when an adjustable-speed motor is desired.

Squirrel-cage and wound-rotor motors are further classified according to the type of enclosure they use to protect their interior windings. These enclosures fall into two broad categories:

- Open
- Totally enclosed

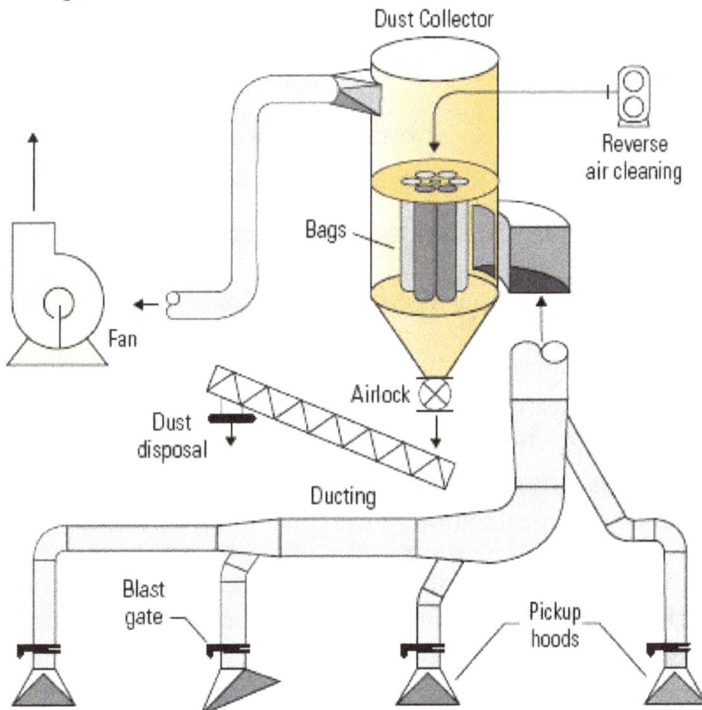

Figure 1. Dust Collection System Example

Drip-proof and splash-proof motors are open motors. They provide varying degrees of protection; however, they should not be used where the air contains substances that might be harmful to the interior of the motor.

Totally enclosed motors are weather-protected with the windings enclosed. These enclosures prevent free exchange of air between the inside and the outside, but they are not airtight.

Totally enclosed, fan-cooled (TEFC) motors are another kind of totally enclosed motor. These motors are the most commonly used motors in dust collection systems. They have an integral-cooling fan outside the enclosure, but within the protective shield, that directs air over the enclosure.

Both open and totally enclosed motors are available in explosion-proof and dust-ignition-proof models to protect against explosion and fire in hazardous environments.

Motors are selected to provide sufficient power to operate fans over the full range of process conditions (temperature and flow rate).

Configurations

Dust collectors can be configured into one of five common types.

1. Ambient units - Ambient units are free-hanging systems for use when applications limit the use of source-capture arms or ductwork.
2. Collection booths - Collector booths require no ductwork, and allow the worker greater freedom of movement. They are often portable.
3. Downdraft tables - A downdraft table is a self-contained portable filtration system that removes harmful particulates and returns filtered air back into the facility with no external ventilation required.
4. Source collector or Portable units - Portable units are for collecting dust, mist, fumes, or smoke at the source.
5. Stationary units - An example of a stationary collector is a baghouse.

Parameters Involved in Specifying Dust Collectors

Important parameters in specifying dust collectors include airflow the velocity of the air stream created by the vacuum producer; system power, the power of the system motor, usually specified in horsepower; storage capacity for dust and particles, and minimum particle size filtered by the unit. Other considerations when choosing a dust collection system include the temperature, moisture content, and the possibility of combustion of the dust being collected.

Systems for fine removal may only contain a single filtration system (such as a filter bag or cartridge). However, most units utilize a primary and secondary separation/filtration system. In many cases the heat or moisture content of dust can negatively affect the filter media of a baghouse or cartridge dust collector. A cyclone separator or dryer may be placed before these units to reduce heat or moisture content before reaching the filters. Furthermore, some units may have third and fourth stage filtration. All separation and filtration systems used within the unit should be specified.

A baghouse is an air pollution abatement device used to trap particulate by filtering gas streams through large fabric bags. They are typically made of glass fibers or fabric.

An impinger system is a device in which particles are removed by impacting the aerosol particles into a liquid. Modular media type units combine a variety of specific filter modules in one unit. These systems can provide solutions to many air contaminant problems. A typical system incorporates a series of disposable or cleanable pre-filters, a disposable vee-bag or cartridge filter. HEPA or carbon final filter modules can also be added. Various models are available, including free-hanging or ducted installations, vertical or horizontal mounting, and fixed or portable configurations. Filter cartridges are made out of a variety of synthetic fibers and are capable of collecting sub-micrometre particles without creating an excessive pressure drop in the system. Filter cartridges require periodic cleaning.

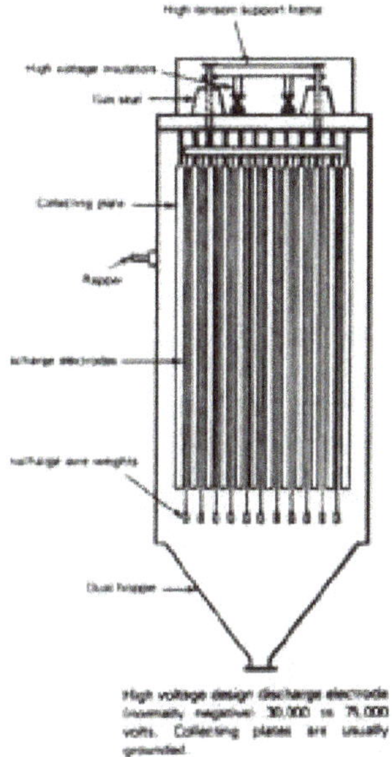

Electrostatic Precipitator

Electrostatic precipitators are a type of air cleaner, which charges particles of dust by passing dust-laden air through a strong (50-100 kV) electrostatic field. This causes the particles to be attracted to oppositely charged plates so that they can be removed from the air stream.

A wet scrubber, or venturi scrubber, is similar to a cyclone but it has an orifice unit that sprays water into the vortex in the cyclone section, collecting all of the dust in a slurry system. The water media can be recirculated and reused to continue to filter the air. Eventually the solids must be removed from the water stream and disposed of.

Filter Cleaning Methods

Online cleaning – automatically timed filter cleaning which allows for continuous, uninterrupted dust collector operation for heavy dust operations.

Offline cleaning – filter cleaning accomplished during dust collector shut down. Practical whenever the dust loading in each dust collector cycle does not exceed the filter capacity. Allows for maximum effectiveness in dislodging and disposing of dust.

On-demand cleaning – filter cleaning initiated automatically when the filter is fully loaded, as determined by a specified drop in pressure across the media surface.

Reverse-pulse/Reverse-jet cleaning – Filter cleaning method which delivers blasts of compressed air from the clean side of the filter to dislodge the accumulated dust cake.

Impact/Rapper cleaning – Filter cleaning method in which high-velocity compressed air forced through a flexible tube results in a random rapping of the filter to dislodge the dust cake. Especially effective when the dust is extremely fine or sticky.

Wet Scrubber

The term wet scrubber describes a variety of devices that remove pollutants from a furnace flue gas or from other gas streams. In a wet scrubber, the polluted gas stream is brought into contact with the scrubbing liquid, by spraying it with the liquid, by forcing it through a pool of liquid, or by some other contact method, so as to remove the pollutants.

Design

The design of wet scrubbers or any air pollution control device depends on the industrial process conditions and the nature of the air pollutants involved. Inlet gas characteristics and dust properties (if particles are present) are of primary importance. Scrubbers can be designed to collect particulate matter and/or gaseous pollutants. The versatility of wet scrubbers allow them to be built in numerous configurations, all designed to provide good contact between the liquid and polluted gas stream.

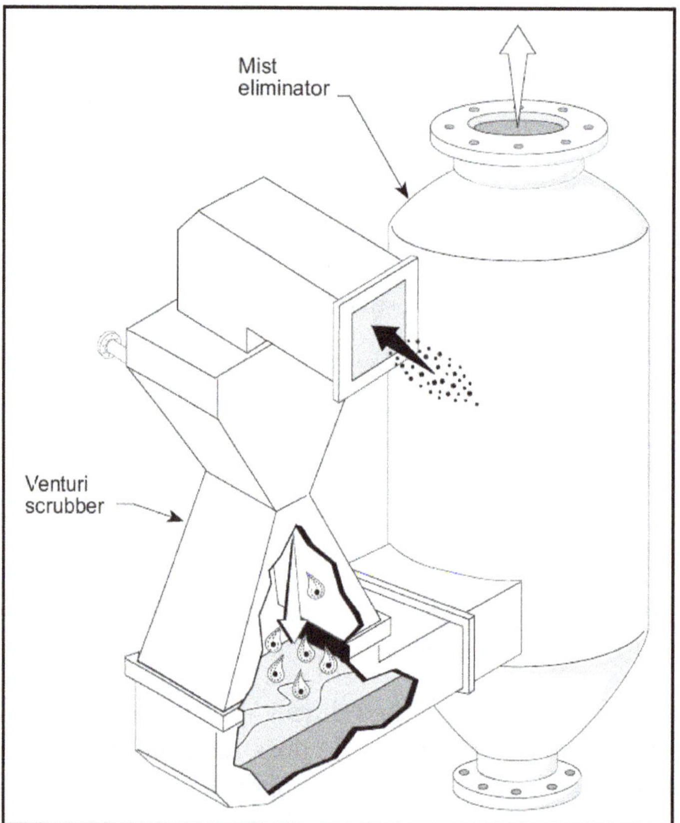

A venturi scrubber design. The mist eliminator for a venturi scrubber is often a separate device called a cyclonic separator

Wet scrubbers remove dust particles by *capturing* them in liquid droplets. The droplets are then collected, the liquid *dissolving* or *absorbing* the pollutant gases. Any droplets that are in the scrubber inlet gas must be separated from the outlet gas stream by means of another device referred to as a mist eliminator or entrainment separator (these terms are interchangeable). Also, the resultant scrubbing liquid must be treated prior to any ultimate discharge or being reused in the plant.

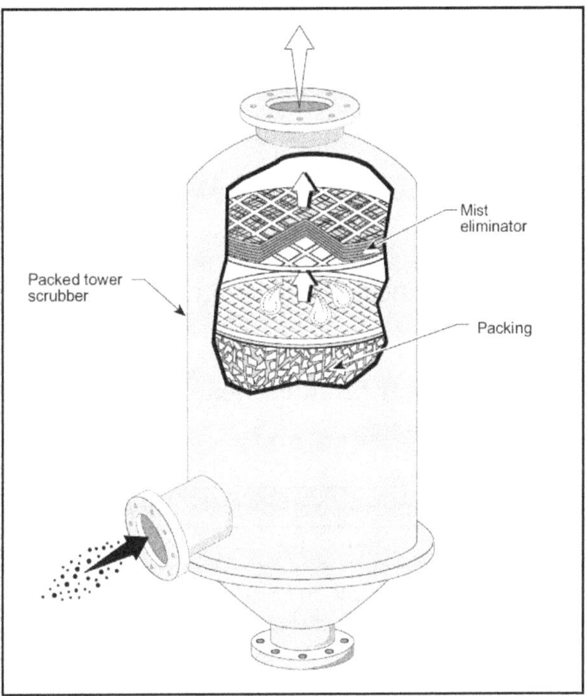

A packed bed tower design where the mist eliminator is built into the top of the structure. Various tower designs exist

A wet scrubber's ability to collect small particles is often directly proportional to the power input into the scrubber. Low energy devices such as spray towers are used to collect particles larger than 5 micrometers. To obtain high efficiency removal of 1 micrometer (or less) particles generally requires high energy devices such as venturi scrubbers or augmented devices such as condensation scrubbers. Additionally, a properly designed and operated entrainment separator or mist eliminator is important to achieve high removal efficiencies. The greater the number of liquid droplets that are not captured by the mist eliminator, the higher the potential emission levels.

Wet scrubbers that remove gaseous pollutants are referred to as *absorbers*. Good gas-to-liquid contact is essential to obtain high removal efficiencies in absorbers. A number of wet scrubber designs are used to remove gaseous pollutants, with the packed tower and the plate tower being the most common.

If the gas stream contains both particle matter and gases, wet scrubbers are generally the only single air pollution control device that can remove both pollutants. Wet scrubbers can achieve high removal efficiencies for either particles or gases and, in some instances, can achieve a high removal efficiency for both pollutants in the same system. However, in many cases, the best operating conditions for particles collection are the poorest for gas removal.

In general, obtaining high simultaneous gas and particulate removal efficiencies requires that one of them be easily collected (i.e., that the gases are very soluble in the liquid or that the particles are large and readily captured) or by the use of a scrubbing reagent such as lime or sodium hydroxide.

Advantages and Disadvantages

For particulate control, wet scrubbers (also referred to as wet collectors) are evaluated against fabric filters and electrostatic precipitators (ESPs). Some *advantages* of wet scrubbers over these devices are as follows:

- Wet scrubbers have the ability to handle high temperatures and moisture.
- In wet scrubbers, flue gases are cooled, resulting in smaller overall size of equipment.
- Wet scrubbers can remove both gases and particulate matter.
- Wet scrubbers can neutralize corrosive gases.

Some *disadvantages* of wet scrubbers include corrosion, the need for entrainment separation or mist removal to obtain high efficiencies and the need for treatment or reuse of spent liquid.

Wet scrubbers have been used in a variety of industries such as acid plants, fertilizer plants, steel mills, asphalt plants, and large power plants.

Relative advantages and disadvantages of wet scrubbers compared to other control devices	
Advantages	**Disadvantages**
• Small space requirements: Scrubbers reduce the temperature and volume of the unsaturated exhaust stream. Therefore, vessel sizes, including fans and ducts downstream, are smaller than those of other control devices. Smaller sizes result in lower capital costs and more flexibility in site location of the scrubber. • No secondary dust sources: Once particulate matter is collected, it cannot escape from hoppers or during transport. • Handles high-temperature, high-humidity gas streams: No temperature limits or condensation problems can occur as in baghouses or ESPs. • Minimal fire and explosion hazards: Various dry dusts are flammable. Using water eliminates the possibility of explosions. • Ability to collect both gases and particulate matter.	• Corrosion problems: Water and dissolved pollutants can form highly corrosive acid solutions. Proper construction materials are very important. Also, wet-dry interface areas can result in corrosion. • High power requirements: High collection efficiencies for particulate matter are attainable only at high pressure drops, resulting in high operating costs. • Water-disposal problems: Settling ponds or sludge clarifiers may be needed to meet waste-water regulations. • Difficult product recovery: Dewatering and drying of scrubber sludge make recovery of any dust for reuse very expensive and difficult.

Components

Wet scrubber systems generally consist of the following components:

- Ductwork and fan system
- A saturation chamber (optional)
- Scrubbing vessel
- Entrainment separator or mist eliminator

- Pumping (and possible recycle system)
- Spent scrubbing liquid treatment and/or reuse system
- An exhaust stack

A typical wet scrubbing process can be described as follows:

- Hot flue gas from a furnace enters a saturator (if present) where gases are cooled and humidified prior to entering the scrubbing area. The saturator removes a small percentage of the particulate matter present in the flue gas.
- Next, the gas enters a venturi scrubber where approximately half of the gases are removed. Venturi scrubbers have a minimum particle removal efficiency of 95%.
- The gas flows through a second scrubber, a packed bed absorber, where the rest of the gases (and particulate matter) are collected.
- An entrainment separator or mist eliminator removes any liquid droplets that may have become entrained in the flue gas.
- A recirculation pump moves some of the spent scrubbing liquid back to the venturi scrubber where it is recycled and the remainder is sent to a treatment system.
- Treated scrubbing liquid is recycled back to the saturator and the packed bed absorber.
- Fans and ductwork move the flue gas stream through the system and eventually out the stack.

Categorization

Since wet scrubbers vary greatly in complexity and method of operation, devising categories into which all of them neatly fit is extremely difficult. Scrubbers for particle collection are usually categorized by the gas-side pressure drop of the system. Gas-side pressure drop refers to the pressure difference, or pressure drop, that occurs as the exhaust gas is pushed or pulled through the scrubber, disregarding the pressure that would be used for pumping or spraying the liquid into the scrubber.

Scrubbers may be classified *by pressure drop* as follows:

- *Low-energy scrubbers* have pressure drops of less than 12.7 cm (5 in) of water.
- *Medium-energy scrubbers* have pressure drops between 12.7 and 38.1 cm (5 and 15 in) of water.
- *High-energy scrubbers* have pressure drops greater than 37.1 cm (15 in) of water.

However, most scrubbers operate over a wide range of pressure drops, depending on their specific application, thereby making this type of categorization difficult.

Another way to classify wet scrubbers is by their *use* - to primarily collect either particulates or gaseous pollutants. Again, this distinction is not always clear since scrubbers can often be used to remove both types of pollutants.

Wet scrubbers can also be categorized by the manner in which the gas and liquid phases are brought into contact. Scrubbers are designed to use power, or energy, from the gas stream or the liquid stream, or some other method to bring the pollutant gas stream into contact with the liquid. These categories are given in Table 2.

Categories of wet collectors by energy source used for contact	
Wet collector	**Energy source used for gas-liquid contact**
• Gas-phase contacting	• Gas stream
• Liquid-phase contacting	• Liquid stream
• Wet film	• Liquid and gas streams
• Combination	• Energy source:
○ Liquid phase and gas phase	○ Liquid and gas streams
○ Mechanically aided	○ Mechanically driven rotor

Material of Construction and Design

Corrosion can be a prime problem associated with chemical industry scrubbing systems. Fibre-reinforced plastic and dual laminates are often used as most dependable materials of construction.

9

Techniques for Air Pollution Mitigation

This chapter on air pollution offers an insightful focus, keeping in mind the complex subject mat-ter. Some techniques to mitigate air pollution are discussed here. Geothermal heat pump and sea-sonal thermal energy storage are two alternative pollution alleviants that this chapter examines. With the help of both, energy can be stored and utilized in a better manner.

Geothermal Heat Pump

A geothermal heat pump or ground source heat pump (GSHP) is a central heating and/or cooling system that transfers heat to or from the ground.

It uses the earth as a heat source (in the winter) or a heat sink (in the summer). This design takes advantage of the moderate temperatures in the ground to boost efficiency and reduce the operational costs of heating and cooling systems, and may be combined with solar heating to form a geosolar system with even greater efficiency. They are also known by other names, including geoexchange, earth-coupled, earth energy systems. The engineering and scientific communities prefer the terms "*geoexchange*" or "*ground source heat pumps*" to avoid confusion with traditional geothermal power, which uses a high temperature heat source to generate electricity. Ground source heat pumps harvest heat absorbed at the Earth's surface from solar energy. The temperature in the ground below 6 metres (20 ft) is roughly equal to the mean annual air temperature at that latitude at the surface.

Depending on latitude, the temperature beneath the upper 6 metres (20 ft) of Earth's surface maintains a nearly constant temperature between 10 and 16 °C (50 and 60 °F), if the temperature is undisturbed by the presence of a heat pump. Like a refrigerator or air conditioner, these systems use a heat pump to force the transfer of heat from the ground. Heat pumps can transfer heat from a cool space to a warm space, against the natural direction of flow, or they can enhance the natural flow of heat from a warm area to a cool one. The core of the heat pump is a loop of refrigerant pumped through a vapor-compression refrigeration cycle that moves heat. Air-source heat pumps are typically more efficient at heating than pure electric heaters, even when extracting heat from cold winter air, although efficiencies begin dropping significantly as outside air temperatures drop below 5 °C (41 °F). A ground source heat pump exchanges heat with the ground. This is much more energy-efficient because underground temperatures are more stable than air temperatures through the year. Seasonal variations drop off with depth and disappear below 7 metres (23 ft) to 12 metres (39 ft) due to thermal inertia. Like a cave, the shallow ground temperature is warmer than the air above during the winter and cooler than the air in the summer. A ground source heat pump extracts ground heat in the winter (for heating) and transfers heat back into the ground in the summer (for cooling). Some systems are designed to operate in one mode only, heating or cooling, depending on climate.

Geothermal pump systems reach fairly high coefficient of performance (CoP), 3 to 6, on the coldest of winter nights, compared to 1.75–2.5 for air-source heat pumps on cool days. Ground source heat pumps (GSHPs) are among the most energy efficient technologies for providing HVAC and water heating.

Setup costs are higher than for conventional systems, but the difference is usually returned in energy savings in 3 to 10 years, and even shorter lengths of time with federal, state and utility tax credits and incentives. Geothermal heat pump systems are reasonably warranted by manufacturers, and their working life is estimated at 25 years for inside components and 50+ years for the ground loop. As of 2004, there are over a million units installed worldwide providing 12 GW of thermal capacity, with an annual growth rate of 10%.

Differing Terms and Definitions

Ground source heating and cooling

Some confusion exists with regard to the terminology of heat pumps and the use of the term "*geothermal*". "*Geothermal*" derives from the Greek and means "*Earth heat*" - which geologists and many laymen understand as describing hot rocks, volcanic activity or heat derived from deep within the earth. Though some confusion arises when the term "*geothermal*" is also used to apply to temperatures within the first 100 metres of the surface, this is "*Earth heat*" all the same, though it is largely influenced by stored energy from the sun.

History

The heat pump was described by Lord Kelvin in 1853 and developed by Peter Ritter von Rittinger in 1855. After experimenting with a freezer, Robert C. Webber built the first direct exchange ground-source heat pump in the late 1940s. The first successful commercial project was installed in the Commonwealth Building (Portland, Oregon) in 1948, and has been designated a National Historic Mechanical Engineering Landmark by ASME. The technology became popular in Sweden in the 1970s, and has been growing slowly in worldwide acceptance since then. Open loop systems dominated the market until the development of polybutylene pipe in 1979 made closed loop systems economically viable. As of 2004, there are over a million units installed worldwide providing 12 GW of thermal capacity. Each year, about 80,000 units are installed in the US (geothermal energy is used in all 50 U.S. states today, with great potential for near-term market growth and savings) and 27,000 in Sweden. In Finland, a geothermal heat pump was the most common heating system choice for new detached houses between 2006 and 2011 with market share exceeding 40%.

Ground Heat Exchanger

Loop field for a 12-ton system (unusually large for most residential applications)

Heat pumps provide winter heating by extracting heat from a source and transferring it into a building. Heat can be extracted from any source, no matter how cold, but a warmer source allows higher efficiency. A ground source heat pump uses the top layer of the earth's crust as a source of heat, thus taking advantage of its seasonally moderated temperature.

In the summer, the process can be reversed so the heat pump extracts heat from the building and transfers it to the ground. Transferring heat to a cooler space takes less energy, so the cooling efficiency of the heat pump gains benefits from the lower ground temperature.

Ground source heat pumps employ a heat exchanger in contact with the ground or groundwater to extract or dissipate heat. This component accounts for anywhere from a fifth to half of the total system cost, and would be the most cumbersome part to repair or replace. Correctly sizing this component is necessary to assure long-term performance: the energy efficiency of the system improves with roughly 4% for every degree Celsius that is won through correct sizing, and the underground temperature balance must be maintained through proper design of the whole system. Incorrect Design can result in the system freezing after a number of years or very inefficient system performance; thus accurate system design is critical to a successful system

Shallow 3–8-foot (0.91–2.44 m) horizontal heat exchangers experience seasonal temperature cycles due to solar gains and transmission losses to ambient air at ground level. These temperature cycles lag behind the seasons because of thermal inertia, so the heat exchanger will harvest heat deposited by the sun several months earlier, while being weighed down in late winter and spring, due to accumulated winter cold. Deep vertical systems 100–500 feet (30–152 m) deep rely on migration of heat from surrounding geology, unless they are recharged annually by solar recharge of the ground or exhaust heat from air conditioning systems.

Several major design options are available for these, which are classified by fluid and layout. Direct exchange systems circulate refrigerant underground, closed loop systems use a mixture of anti-freeze and water, and open loop systems use natural groundwater.

Direct Exchange

The direct exchange geothermal heat pump is the oldest type of geothermal heat pump technology. The ground-coupling is achieved through a single loop, circulating refrigerant, in direct thermal contact with the ground (as opposed to a combination of a refrigerant loop and a water loop). The refrigerant leaves the heat pump cabinet, circulates through a loop of copper tube buried underground, and exchanges heat with the ground before returning to the pump. The name "direct exchange" refers to heat transfer between the refrigerant loop and the ground without the use of an intermediate fluid. There is no direct interaction between the fluid and the earth; only heat transfer through the pipe wall. Direct exchange heat pumps, which are now rarely used, are not to be confused with "water-source heat pumps" or "water loop heat pumps" since there is no water in the ground loop. ASHRAE defines the term ground-coupled heat pump to encompass closed loop and direct exchange systems, while excluding open loops.

Direct exchange systems are more efficient and have potentially lower installation costs than closed loop water systems. Copper's high thermal conductivity contributes to the higher efficiency of the system, but heat flow is predominantly limited by the thermal conductivity of the ground, not the pipe. The main reasons for the higher efficiency are the elimination of the water pump (which uses electricity), the elimination of the water-to-refrigerant heat exchanger (which is a source of heat losses), and most importantly, the latent heat phase change of the refrigerant in the ground itself.

While they require more refrigerant and their tubing is more expensive per foot, a direct exchange earth loop is shorter than a closed water loop for a given capacity. A direct exchange system requires only 15 to 30% of the length of tubing and half the diameter of drilled holes, and the drilling or excavation costs are therefore lower. Refrigerant loops are less tolerant of leaks than water loops because gas can leak out through smaller imperfections. This dictates the use of brazed copper tubing, even though the pressures are similar to water loops. The copper loop must be protected from corrosion in acidic soil through the use of a sacrificial anode or other cathodic protection.

The U.S. Environmental Protection Agency conducted field monitoring of a direct geoexchange heat pump water heating system in a commercial application. The EPA reported that the system saved 75% of the electrical energy that would have been required by an electrical resistance water heating unit. According to the EPA, if the system is operated to capacity, it can avoid the emission of up to 7,100 pounds of CO_2 and 15 pounds of NO_x each year per ton of compressor capacity (or 42,600 lbs. of CO_2 and 90 lbs. of NO_x for a typical 6 ton system).

In Northern climates, although the earth temperature is cooler, so is the incoming water temperature, which enables the high efficiency systems to replace more energy than would otherwise be required of electric or fossil fuel fired systems. Any temperature above -40 °F is sufficient to evaporate the refrigerant, and the direct exchange system can harvest energy through ice.

In extremely hot climates with dry soil, the addition of an auxiliary cooling module as a second condenser in line between the compressor and the earth loops increases efficiency and can further reduce the amount of earth loop to be installed.

Closed Loop

Most installed systems have two loops on the ground side: the primary refrigerant loop is contained in the appliance cabinet where it exchanges heat with a secondary water loop that is buried underground. The secondary loop is typically made of high-density polyethylene pipe and contains a mixture of water and anti-freeze (propylene glycol, denatured alcohol or methanol). Monopropylene glycol has the least damaging potential when it might leak into the ground, and is therefore the only allowed anti-freeze in ground sources in an increasing number of European countries. After leaving the internal heat exchanger, the water flows through the secondary loop outside the building to exchange heat with the ground before returning. The secondary loop is placed below the frost line where the temperature is more stable, or preferably submerged in a body of water if available. Systems in wet ground or in water are generally more efficient than drier ground loops since water conducts and stores heat better than solids in sand or soil. If the ground is naturally dry, soaker hoses may be buried with the ground loop to keep it wet.

An installed liquid pump pack

Closed loop systems need a heat exchanger between the refrigerant loop and the water loop, and pumps in both loops. Some manufacturers have a separate ground loop fluid pump pack, while some integrate the pumping and valving within the heat pump. Expansion tanks and pressure relief valves may be installed on the heated fluid side. Closed loop systems have lower efficiency than direct exchange systems, so they require longer and larger pipe to be placed in the ground, increasing excavation costs.

Closed loop tubing can be installed horizontally as a loop field in trenches or vertically as a series of long U-shapes in wells. The size of the loop field depends on the soil type and moisture content, the average ground temperature and the heat loss and or gain characteristics of the building being conditioned. A rough approximation of the initial soil temperature is the average daily temperature for the region.

Vertical

Drilling of a borehole for residential heating

A vertical closed loop field is composed of pipes that run vertically in the ground. A hole is bored in the ground, typically 50 to 400 feet (15–122 m) deep. Pipe pairs in the hole are joined with a U-shaped cross connector at the bottom of the hole. The borehole is commonly filled with a bentonite grout surrounding the pipe to provide a thermal connection to the surrounding soil or rock to improve the heat transfer. Thermally enhanced grouts are available to improve this heat transfer. Grout also protects the ground water from contamination, and prevents artesian wells from flooding the property. Vertical loop fields are typically used when there is a limited area of land available. Bore holes are spaced at least 5–6 m apart and the depth depends on ground and building characteristics. For illustration, a detached house needing 10 kW (3 ton) of heating capacity might need three boreholes 80 to 110 m (260 to 360 ft) deep. (A ton of heat is 12,000 British thermal units per hour (BTU/h) or 3.5 kilowatts.) During the cooling season, the local temperature rise in the bore field is influenced most by the moisture travel in the soil. Reliable heat transfer models have been developed through sample bore holes as well as other tests.

Horizontal

A horizontal closed loop field is composed of pipes that run horizontally in the ground. A long horizontal trench, deeper than the frost line, is dug and U-shaped or slinky coils are placed horizontally inside the same trench. Excavation for shallow horizontal loop fields is about half the cost of vertical drilling, so this is the most common layout used wherever there is adequate land available. For illustration, a detached house needing 10 kW (3 ton) of heating capacity might need three loops 120 to 180 m (390 to 590 ft) long of NPS 3/4 (DN 20) or NPS 1.25 (DN 32) polyethylene tubing at a depth of 1 to 2 m (3.3 to 6.6 ft).

A three-ton slinky loop prior to being covered with soil. The three slinky loops are running out horizontally with three straight lines returning the end of the slinky coil to the heat pump.

The depth at which the loops are placed significantly influences the energy consumption of the heat pump in two opposite ways: shallow loops tend to indirectly absorb more heat from the sun, which is helpful, especially when the ground is still cold after a long winter. On the other hand, shallow loops are also cooled down much more readily by weather changes, especially during long cold winters, when heating demand peaks. Often, the second effect is much greater than the first one, leading to higher costs of operation for the more shallow ground loops. This problem can be reduced by increasing both the depth and the length of piping, thereby significantly increasing costs of installation. However, such expenses might be deemed feasible, as they may result in lower operating costs. Recent studies show that utilization of a non-homogeneous soil profile with a layer of low conductive material above the ground pipes can help mitigate the adverse effects of shallow pipe burial depth. The intermediate blanket with lower conductivity than the surrounding soil profile demonstrated the potential to increase the energy extraction rates from the ground to as high as 17% for a cold climate and about 5-6% for a relatively moderate climate.

A slinky (also called coiled) closed loop field is a type of horizontal closed loop where the pipes overlay each other (not a recommended method). The easiest way of picturing a slinky field is to imagine holding a slinky on the top and bottom with your hands and then moving your hands in opposite directions. A slinky loop field is used if there is not adequate room for a true horizontal system, but it still allows for an easy installation. Rather than using straight pipe, slinky coils use overlapped loops of piping laid out horizontally along the bottom of a wide trench. Depending on soil, climate and the heat pump's run fraction, slinky coil trenches can be up to two thirds shorter than traditional horizontal loop trenches. Slinky coil ground loops are essentially a more economical and space efficient version of a horizontal ground loop.

Radial or Directional Drilling

As an alternative to trenching, loops may be laid by mini horizontal directional drilling (mini-HDD). This technique can lay piping under yards, driveways, gardens or other structures without disturbing them, with a cost between those of trenching and vertical drilling. This system also differs from horizontal & vertical drilling as the loops are installed from one central chamber, further reducing the ground space needed. Radial drilling is often installed retroactively (after the property has been built) due to the small nature of the equipment used and the ability to bore beneath existing constructions.

Pond

12-ton pond loop system being sunk to the bottom of a pond

A closed pond loop is not common because it depends on proximity to a body of water, where an open loop system is usually preferable. A pond loop may be advantageous where poor water quality precludes an open loop, or where the system heat load is small. A pond loop consists of coils of pipe similar to a slinky loop attached to a frame and located at the bottom of an appropriately sized pond or water source.

Open Loop

In an open loop system (also called a groundwater heat pump), the secondary loop pumps natural water from a well or body of water into a heat exchanger inside the heat pump. ASHRAE calls open loop systems *groundwater heat pumps* or *surface water heat pumps*, depending on the source. Heat is either extracted or added by the primary refrigerant loop, and the water is returned to a separate injection well, irrigation trench, tile field or body of water. The supply and return lines must be placed far enough apart to ensure thermal recharge of the source. Since the water chemistry is not controlled, the appliance may need to be protected from corrosion by using different metals in the heat exchanger and pump. Limescale may foul the system over time and require periodic acid cleaning. This is much more of a problem with cooling systems than heating systems. Also, as fouling decreases the flow of natural water, it becomes difficult for the heat pump to exchange building heat with the groundwater. If the water contains high levels of salt, minerals, iron bacteria or hydrogen sulfide, a closed loop system is usually preferable.

Deep lake water cooling uses a similar process with an open loop for air conditioning and cooling. Open loop systems using ground water are usually more efficient than closed systems because they are better coupled with ground temperatures. Closed loop systems, in comparison, have to transfer heat across extra layers of pipe wall and dirt.

A growing number of jurisdictions have outlawed open-loop systems that drain to the surface because these may drain aquifers or contaminate wells. This forces the use of more environmentally sound injection wells or a closed loop system.

Standing Column Well

A standing column well system is a specialized type of open loop system. Water is drawn from the bottom of a deep rock well, passed through a heat pump, and returned to the top of the well, where traveling downwards it exchanges heat with the surrounding bedrock. The choice of a standing column well system is often dictated where there is near-surface bedrock and limited surface area is available. A standing column is typically not suitable in locations where the geology is mostly clay, silt, or sand. If bedrock is deeper than 200 feet (61 m) from the surface, the cost of casing to seal off the overburden may become prohibitive.

A multiple standing column well system can support a large structure in an urban or rural application. The standing column well method is also popular in residential and small commercial applications. There are many successful applications of varying sizes and well quantities in the many boroughs of New York City, and is also the most common application in the New England states. This type of ground source system has some heat storage benefits, where heat is rejected from the building and the temperature of the well is raised, within reason, during the summer cooling months which can then be harvested for heating in the winter months, thereby increasing the efficiency of the heat pump system. As with closed loop systems, sizing of the standing column system is critical in reference to the heat loss and gain of the existing building. As the heat exchange is actually with the bedrock, using water as the transfer medium, a large amount of production capacity (water flow from the well) is not required for a standing column system to work. However, if there is adequate water production, then the thermal capacity of the well system can be enhanced by discharging a small percentage of system flow during the peak Summer and Winter months.

Since this is essentially a water pumping system, standing column well design requires critical considerations to obtain peak operating efficiency. Should a standing column well design be misapplied, leaving out critical shut-off valves for example, the result could be an extreme loss in efficiency and thereby cause operational cost to be higher than anticipated.

Building Distribution

The heat pump is the central unit that becomes the heating and cooling plant for the building. Some models may cover space heating, space cooling, (space heating via conditioned air, hydronic systems and / or radiant heating systems), domestic or pool water preheat (via the desuperheater function), demand hot water, and driveway ice melting all within one appliance with a variety of options with respect to controls, staging and zone control. The heat may be carried to its end use by circulating water or forced air. Almost all types of heat pumps are produced for commercial and residential applications.

Liquid-to-air heat pump

Liquid-to-air heat pumps (also called *water-to-air*) output forced air, and are most commonly used to replace legacy forced air furnaces and central air conditioning systems. There are variations that allow for split systems, high-velocity systems, and ductless systems. Heat pumps cannot achieve as high a fluid temperature as a conventional furnace, so they require a higher volume flow rate of air to compensate. When retrofitting a residence, the existing duct work may have to be enlarged to reduce the noise from the higher air flow.

Liquid-to-water heat pumps (also called *water-to-water*) are hydronic systems that use water to carry heating or cooling through the building. Systems such as radiant underfloor heating, baseboard radiators, conventional cast iron radiators would use a liquid-to-water heat pump. These heat pumps are preferred for pool heating or domestic hot water pre-heat. Heat pumps can only heat water to about 50 °C (122 °F) efficiently, whereas a boiler normally reaches 65–95 °C (149–203 °F). Legacy radiators designed for these higher temperatures may have to be doubled in numbers when retrofitting a home. A hot water tank will still be needed to raise water temperatures above the heat pump's maximum, but pre-heating will save 25–50% of hot water costs.

Ground source heat pumps are especially well matched to underfloor heating and baseboard radiator systems which only require warm temperatures 40 °C (104 °F) to work well. Thus they are ideal for open plan offices. Using large surfaces such as floors, as opposed to radiators, distributes the heat more uniformly and allows for a lower water temperature. Wood or carpet floor coverings dampen this effect because the thermal transfer efficiency of these materials is lower than that of masonry floors (tile, concrete). Underfloor piping, ceiling or wall radiators can also be used for cooling in dry climates, although the temperature of the circulating water must be above the dew point to ensure that atmospheric humidity does not condense on the radiator.

Liquid-to-water heat pump

Combination heat pumps are available that can produce forced air and circulating water simultaneously and individually. These systems are largely being used for houses that have a combination of air and liquid conditioning needs, for example central air conditioning and pool heating.

Seasonal Thermal Storage

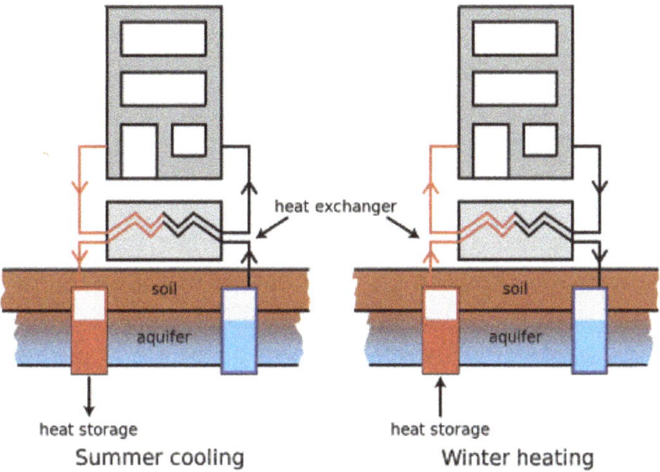

A heat pump in combination with heat and cold storage

The efficiency of ground source heat pumps can be greatly improved by using seasonal thermal energy storage and interseasonal heat transfer. Heat captured and stored in thermal banks in the summer can be retrieved efficiently in the winter. Heat storage efficiency increases with scale, so this advantage is most significant in commercial or district heating systems.

Geosolar combisystems have been used to heat and cool a greenhouse using an aquifer for thermal storage. In summer, the greenhouse is cooled with cold ground water. This heats the water in the aquifer which can become a warm source for heating in winter. The combination of cold and heat storage with heat pumps can be combined with water/humidity regulation. These principles are used to provide renewable heat and renewable cooling to all kinds of buildings.

Also the efficiency of existing small heat pump installations can be improved by adding large, cheap, water filled solar collectors. These may be integrated into a to-be-overhauled parking lot, or in walls or roof constructions by installing one inch PE pipes into the outer layer.

Thermal Efficiency

The net thermal efficiency of a heat pump should take into account the efficiency of electricity generation and transmission, typically about 30%. Since a heat pump moves three to five times more heat energy than the electric energy it consumes, the total energy output is much greater than the electrical input. This results in net thermal efficiencies greater than 300% as compared to radiant electric heat being 100% efficient. Traditional combustion furnaces and electric heaters can never exceed 100% efficiency.

Geothermal heat pumps can reduce energy consumption— and corresponding air pollution emissions—up to 44% compared to air source heat pumps and up to 72% compared to electric resistance heating with standard air-conditioning equipment.

The dependence of net thermal efficiency on the electricity infrastructure tends to be an unnecessary complication for consumers and is not applicable to hydroelectric power, so performance of heat pumps is usually expressed as the ratio of heating output or heat removal to electricity input. Cooling performance is typically expressed in units of BTU/hr/watt as the energy efficiency ratio (EER), while heating performance is typically reduced to dimensionless units as the coefficient of performance (COP). The conversion factor is 3.41 BTU/hr/watt. Performance is influenced by all components of the installed system, including the soil conditions, the ground-coupled heat exchanger, the heat pump appliance, and the building distribution, but is largely determined by the "lift" between the input temperature and the output temperature.

For the sake of comparing heat pump appliances to each other, independently from other system components, a few standard test conditions have been established by the American Refrigerant Institute (ARI) and more recently by the International Organization for Standardization. Standard ARI 330 ratings were intended for closed loop ground-source heat pumps, and assume secondary loop water temperatures of 77 °F (25 °C) for air conditioning and 32 °F (0 °C) for heating. These temperatures are typical of installations in the northern US. Standard ARI 325 ratings were intended for open loop ground-source heat pumps, and include two sets of ratings for groundwater temperatures of 50 °F (10 °C) and 70 °F (21 °C). ARI 325 budgets more electricity for water pumping than ARI 330. Neither of these standards attempt to account for seasonal variations. Standard ARI 870 ratings are intended for direct exchange ground-source heat pumps. ASHRAE transitioned to ISO 13256-1 in 2001, which replaces ARI 320, 325 and 330. The new ISO standard produces slightly higher ratings because it no longer budgets any electricity for water pumps.

Efficient compressors, variable speed compressors and larger heat exchangers all contribute to heat pump efficiency. Residential ground source heat pumps on the market today have standard COPs ranging from 2.4 to 5.0 and EERs ranging from 10.6 to 30. To qualify for an Energy Star label, heat pumps must meet certain minimum COP and EER ratings which depend on the ground heat exchanger type. For closed loop systems, the ISO 13256-1 heating COP must be 3.3 or greater and the cooling EER must be 14.1 or greater.

Actual installation conditions may produce better or worse efficiency than the standard test conditions. COP improves with a lower temperature difference between the input and output of the heat pump, so the stability of ground temperatures is important. If the loop field or water pump is undersized, the addition or removal of heat may push the ground temperature beyond standard test conditions, and performance will be degraded. Similarly, an undersized blower may allow the plenum coil to overheat and degrade performance.

Soil without artificial heat addition or subtraction and at depths of several metres or more remains at a relatively constant temperature year round. This temperature equates roughly to the average annual air-temperature of the chosen location, usually 7–12 °C (45–54 °F) at a depth of 6 metres (20 ft) in the northern US. Because this temperature remains more constant than the air temperature throughout the seasons, geothermal heat pumps perform with far greater efficiency during extreme air temperatures than air conditioners and air-source heat pumps.

Standards ARI 210 and 240 define Seasonal Energy Efficiency Ratio (SEER) and Heating Seasonal Performance Factors (HSPF) to account for the impact of seasonal variations on air source heat pumps. These numbers are normally not applicable and should not be compared to ground source heat pump ratings. However, Natural Resources Canada has adapted this approach to calculate typical seasonally adjusted HSPFs for ground-source heat pumps in Canada. The NRC HSPFs ranged from 8.7 to 12.8 BTU/hr/watt (2.6 to 3.8 in nondimensional factors, or 255% to 375% seasonal average electricity utilization efficiency) for the most populated regions of Canada. When combined with the thermal efficiency of electricity, this corresponds to net average thermal efficiencies of 100% to 150%.

Environmental Impact

The US Environmental Protection Agency (EPA) has called ground source heat pumps the most energy-efficient, environmentally clean, and cost-effective space conditioning systems available. Heat pumps offer significant emission reductions potential, particularly where they are used for both heating and cooling and where the electricity is produced from renewable resources.

Ground-source heat pumps have unsurpassed thermal efficiencies and produce zero emissions locally, but their electricity supply includes components with high greenhouse gas emissions, unless the owner has opted for a 100% renewable energy supply. Their environmental impact therefore depends on the characteristics of the electricity supply and the available alternatives.

Annual greenhouse gas (GHG) savings from using a ground source heat pump instead of a high-efficiency furnace in a detached residence (assuming no specific supply of renewable energy)				
Country	**Electricity CO_2 Emissions Intensity**	**GHG savings relative to**		
		natural gas	**heating oil**	**electric heating**
Canada	223 ton/GWh	2.7 ton/yr	5.3 ton/yr	3.4 ton/yr
Russia	351 ton/GWh	1.8 ton/yr	4.4 ton/yr	5.4 ton/yr
US	676 ton/GWh	-0.5 ton/yr	2.2 ton/yr	10.3 ton/yr
China	839 ton/GWh	-1.6 ton/yr	1.0 ton/yr	12.8 ton/yr

The GHG emissions savings from a heat pump over a conventional furnace can be calculated based on the following formula:

$$GHG\ Savings = HL\left(\frac{FI}{AFUE \times 1000\frac{kg}{ton}} - \frac{EI}{COP \times 3600\frac{sec}{hr}}\right)$$

- HL = seasonal heat load ≈ 80 GJ/yr for a modern detached house in the northern US
- FI = emissions intensity of fuel = 50 kg(CO_2)/GJ for natural gas, 73 for heating oil, 0 for 100% renewable energy such as wind, hydro, photovoltaic or solar thermal
- AFUE = furnace efficiency ≈ 95% for a modern condensing furnace
- COP = heat pump coefficient of performance ≈ 3.2 seasonally adjusted for northern US heat pump
- EI = emissions intensity of electricity ≈ 200-800 ton(CO_2)/GWh, depending on region

Ground-source heat pumps always produce fewer greenhouse gases than air conditioners, oil furnaces, and electric heating, but natural gas furnaces may be competitive depending on the greenhouse gas intensity of the local electricity supply. In countries like Canada and Russia with low emitting electricity infrastructure, a residential heat pump may save 5 tons of carbon dioxide per year relative to an oil furnace, or about as much as taking an average passenger car off the road. But in cities like Beijing or Pittsburgh that are highly reliant on coal for electricity production, a heat pump may result in 1 or 2 tons more carbon dioxide emissions than a natural gas furnace. For areas not served by utility natural gas infrastructure, however, no better alternative exists.

The fluids used in closed loops may be designed to be biodegradable and non-toxic, but the refrigerant used in the heat pump cabinet and in direct exchange loops was, until recently, chlorodifluoromethane, which is an ozone depleting substance. Although harmless while contained, leaks and improper end-of-life disposal contribute to enlarging the ozone hole. For new construction, this refrigerant is being phased out in favor of the ozone-friendly but potent greenhouse gas R410A. The EcoCute water heater is an air-source heat pump that uses carbon dioxide as its working fluid instead of chlorofluorocarbons. Open loop systems (i.e. those that draw ground water as opposed to closed loop systems using a borehole heat exchanger) need to be balanced by reinjecting the spent water. This prevents aquifer depletion and the contamination of soil or surface water with brine or other compounds from underground.

Before drilling the underground geology needs to be understood, and drillers need to be prepared to seal the borehole, including preventing penetration of water between strata. The unfortunate example is a geothermal heating project in Staufen im Breisgau, Germany which seems the cause of considerable damage to historical buildings there. In 2008, the city centre was reported to have risen 12 cm, after initially sinking a few millimeters. The boring tapped a naturally pressurized aquifer, and via the borehole this water entered a layer of anhydrite, which expands when wet as it forms gypsum. The swelling will stop when the anhydrite is fully reacted, and reconstruction of the city center "is not expedient until the uplift ceases." By 2010 sealing of the borehole had not been accomplished. By 2010, some sections of town had risen by 30 cm.

Ground-source heat pump technology, like building orientation, is a natural building technique (bioclimatic building).

Economics

Ground source heat pumps are characterized by high capital costs and low operational costs compared to other HVAC systems. Their overall economic benefit depends primarily on the relative costs of electricity and fuels, which are highly variable over time and across the world. Based on recent prices, ground-source heat pumps currently have lower operational costs than any other conventional heating source almost everywhere in the world. Natural gas is the only fuel with competitive operational costs, and only in a handful of countries where it is exceptionally cheap, or where electricity is exceptionally expensive. In general, a homeowner may save anywhere from 20% to 60% annually on utilities by switching from an ordinary system to a ground-source system.

Capital costs and system lifespan have received much less study until recently, and the return on investment is highly variable. The most recent data from an analysis of 2011-2012 incentive payments in the state of Maryland showed an average cost of residential systems of $1.90 per watt, or about $26,700 for a typical (4 ton) home system. An older study found the total installed cost for a system with 10 kW (3 ton) thermal capacity for a detached rural residence in the US averaged $8000–$9000 in 1995 US dollars. More recent studies found an average cost of $14,000 in 2008 US dollars for the same size system. The US Department of Energy estimates a price of $7500 on its website, last updated in 2008. One source in Canada placed prices in the range of $30,000-$34,000 Canadian dollars. The rapid escalation in system price has been accompanied by rapid improvements in efficiency and reliability. Capital costs are known to benefit from economies of scale, particularly for open loop systems, so they are more cost-effective for larger commercial buildings and harsher climates. The initial cost can be two to five times that of a conventional heating system in most residential applications, new construction or existing. In retrofits, the cost of installation is affected by the size of living area, the home's age, insulation characteristics, the geology of the area, and location of the property. Proper duct system design and mechanical air exchange should be considered in the initial system cost.

Payback period for installing a ground source heat pump in a detached residence

Country	Payback period for replacing		
	natural gas	**heating oil**	**electric heating**
Canada	13 years	3 years	6 years
US	12 years	5 years	4 years

Germany	net loss	8 years	2 years
Notes: • Highly variable with energy prices. • Government subsidies not included. • Climate differences not evaluated.			

Capital costs may be offset by government subsidies; for example, Ontario offered $7000 for residential systems installed in the 2009 fiscal year. Some electric companies offer special rates to customers who install a ground-source heat pump for heating or cooling their building. Where electrical plants have larger loads during summer months and idle capacity in the winter, this increases electrical sales during the winter months. Heat pumps also lower the load peak during the summer due to the increased efficiency of heat pumps, thereby avoiding costly construction of new power plants. For the same reasons, other utility companies have started to pay for the installation of ground-source heat pumps at customer residences. They lease the systems to their customers for a monthly fee, at a net overall saving to the customer.

The lifespan of the system is longer than conventional heating and cooling systems. Good data on system lifespan is not yet available because the technology is too recent, but many early systems are still operational today after 25–30 years with routine maintenance. Most loop fields have warranties for 25 to 50 years and are expected to last at least 50 to 200 years. Ground-source heat pumps use electricity for heating the house. The higher investment above conventional oil, propane or electric systems may be returned in energy savings in 2–10 years for residential systems in the US. If compared to natural gas systems, the payback period can be much longer or non-existent. The payback period for larger commercial systems in the US is 1–5 years, even when compared to natural gas. Additionally, because geothermal heat pumps usually have no outdoor compressors or cooling towers, the risk of vandalism is reduced or eliminated, potentially extending a system's lifespan.

Ground source heat pumps are recognized as one of the most efficient heating and cooling systems on the market. They are often the second-most cost effective solution in extreme climates (after co-generation), despite reductions in thermal efficiency due to ground temperature. (The ground source is warmer in climates that need strong air conditioning, and cooler in climates that need strong heating.)

Commercial systems maintenance costs in the US have historically been between $0.11 to $0.22 per m^2 per year in 1996 dollars, much less than the average $0.54 per m^2 per year for conventional HVAC systems.

Governments that promote renewable energy will likely offer incentives for the consumer (residential), or industrial markets. For example, in the United States, incentives are offered both on the state and federal levels of government. In the United Kingdom the Renewable Heat Incentive provides a financial incentive for generation of renewable heat based on metered readings on an annual basis for 20 years for commercial buildings. The domestic Renewable Heat Incentive is due to be introduced in Spring 2014 for seven years and be based on deemed heat.

Installation

Because of the technical knowledge and equipment needed to design and size the system properly (and install the piping if heat fusion is required), a GSHP system installation requires a professional's services. Several installers have published real-time views of system performance in an online community of recent residential installations. The International Ground Source Heat Pump Association (IGSHPA), Geothermal Exchange Organization (GEO), the Canadian GeoExchange Coalition and the Ground Source Heat Pump Association maintain listings of qualified installers in the US, Canada and the UK. Furthermore, detailed analysis of Soil thermal conductivity for horizontal systems and formation thermal conductivity for vertical systems will generally result in more accurately design systems with a higher efficiency.

Seasonal Thermal Energy Storage

Seasonal thermal energy storage (or STES) is the storage of heat or cold for periods of up to several months. The thermal energy can be collected whenever it is available and be used whenever needed, such as in the opposing season. For example, heat from solar collectors or waste heat from air conditioning equipment can be gathered in hot months for space heating use when needed, including during winter months. Waste heat from industrial process can similarly be stored and be used much later. Or the natural cold of winter air can be stored for summertime air conditioning. STES stores can serve district heating systems, as well as single buildings or complexes. Among seasonal storages used for heating, the design peak annual temperatures generally are in the range of 27 to 80 °C (80.6 to 176.0 °F), and the temperature difference occurring in the storage over the course of a year can be several tens of degrees. Some systems use a heat pump to help charge and discharge the storage during part or all of the cycle. For cooling applications, often only circulation pumps are used. A less common term for STES technologies is interseasonal thermal energy storage

Examples for district heating include Drake Landing Solar Community where ground storage provide 97% of yearly consumption without heat pumps, and Danish pond storage with boosting.

STES Technologies

There are several types of STES technology, covering a range of applications from single small buildings to community district heating networks. Generally, efficiency increases and the specific construction cost decreases with size.

Underground Thermal Energy Storage

- UTES (underground thermal energy storage), in which the storage medium may be geological strata ranging from earth or sand to solid bedrock, or aquifers. UTES technologies include:
    - ATES (aquifer thermal energy storage). An ATES store is composed of a doublet, totaling two or more wells into a deep aquifer that is contained between impermeable

geological layers above and below. One half of the doublet is for water extraction and the other half for reinjection, so the aquifer is kept in hydrological balance, with no net extraction. The heat (or cold) storage medium is the water and the substrate it occupies. Germany's Reichstag building has been both heated and cooled since 1999 with ATES stores, in two aquifers at different depths. In the Netherlands there are now well over 1,000 ATES systems, which are now a standard construction option. A significant system has been operating at Richard Stockton College (New Jersey) for several years. ATES has a lower installation cost than BTES because usually fewer holes are drilled, but ATES has a higher operating cost. Also, ATES requires particular underground conditions to be feasible, including the presence of an aquifer.

- BTES (borehole thermal energy storage). BTES stores can be constructed wherever boreholes can be drilled, and are composed of one to hundreds of vertical boreholes, typically 155 mm (6.102 in) in diameter. Systems of all sizes have been built, including many quite large. The strata can be anything from sand to crystalline hardrock, and depending on engineering factors the depth can be from 50 to 300 metres (164 to 984 ft). Spacings have ranged from 3 to 8 metres (9.8 to 26.2 ft). Thermal models can be used to predict seasonal temperature variation in the ground, including the establishment of a stable temperature regime which is achieved by matching the inputs and outputs of heat over one or more annual cycles. Warm-temperature seasonal heat stores can be created using borehole fields to store surplus heat captured in summer to actively raise the temperature of large thermal banks of soil so that heat can be extracted more easily (and more cheaply) in winter. Interseasonal Heat Transfer uses water circulating in pipes embedded in asphalt solar collectors to transfer heat to Thermal Banks created in borehole fields. A ground source heat pump is used in winter to extract the warmth from the Thermal Bank to provide space heating via underfloor heating. A high Coefficient of Performance is obtained because the heat pump starts with a warm temperature of 25 °C (77 °F) from the thermal store, instead of a cold temperature of 10 °C (50 °F) from the ground. A BTES operating at Richard Stockton College since 1995 at a peak of about 29 °C (84.2 °F) consists of 400 boreholes 130 metres (427 ft) deep under a 3.5-acre (1.4 ha) parking lot. It has a heat loss of 2% over six months. The upper temperature limit for a BTES store is 85 °C (185 °F) due to characteristics of the PEX pipe used for BHEs, but most do not approach that limit. Boreholes can be either grout- or water-filled depending on geological conditions, and usually have a life expectancy in excess of 100 years. Both a BTES and its associated district heating system can be expanded incrementally after operation begins, as at Neckarsulm, Germany. BTES stores generally do not impair use of the land, and can exist under buildings, agricultural fields and parking lots. An example of one of the several kinds of STES illustrates well the capability of interseasonal heat storage. In Alberta, Canada, the homes of the Drake Landing Solar Community (in operation since 2007), get 97% of their year-round heat from a district heat system that is supplied by solar heat from solar-thermal panels on garage roofs. This feat – a world record – is enabled by interseasonal heat storage in a large mass of native rock that is under a central park. The thermal exchange occurs via a cluster of 144 boreholes, drilled 37 metres

(121 ft) into the earth. Each borehole is 155 mm (6.1 in) in diameter and contains a simple heat exchanger made of small diameter plastic pipe, through which water is circulated. No heat pumps are involved.

- CTES (cavern or mine thermal energy storage). STES stores are possible in flooded mines, purpose-built chambers, or abandoned underground oil stores (e.g. those mined into crystalline hardrock in Norway), if they are close enough to a heat (or cold) source and market.
- Energy Pilings. During construction of large buildings, BHE heat exchangers much like those used for BTES stores have been spiraled inside the cages of reinforcement bars for pilings, with concrete then poured in place. The pilings and surrounding strata then become the storage medium.

Surface and Above Ground Technologies

- Pit storages. Lined, shallow dug pits that are filled with gravel and water as the storage medium are used for STES in many Danish district heating systems. Pit storages are covered with a layer of insulation and then soil, and are used for agriculture or other purposes. Marstal, Denmark's system is a case study, initially providing 20% of the village's year-round heat but now being expanded to provide twice that. The worlds largest pit store (200,000 m3) was commissioned in Denmark in 2015, and allows solar heat to provide 50% of the annual energy for the world's largest solar-enabled district heating system.
- Large-scale water storages. Large scale STES water storage tanks can be built above ground, insulated, and then covered with soil.
- Horizontal heat exchangers. For small installations, a "slinky" heat exchanger of plastic pipe can be shallow-buried in a trench to create an STES.
- Earth-bermed buildings, with passive heat storage in surrounding soil.

Conferences and Organizations

The International Energy Agency's *Energy Conservation through Energy Storage (ECES) Programme* has held triennial global energy conferences since 1981. The conferences originally focused exclusively on STES, but now that those technologies are mature other topics such as phase change materials (PCM) and electrical energy storage are also being covered. Since 1985 each conference has had "stock" (for storage) at the end of its name; e.g. Ecostock, Thermastock. They are held at various locations around the world. Most recent was Innostock 2012 (the 12th International Conference on Thermal Energy Storage) in Lleida, Spain. Greenstock 2015 will be held in Beijing.

The IEA-ECES programme continues the work of the earlier *International Council for Thermal Energy Storage* which from 1978 to 1990 had a quarterly newsletter and was initially sponsored by the U.S. Department of Energy. The newsletter was initially called *ATES Newsletter*, and after BTES became a feasible technology it was changed to *STES Newsletter*.

Use of STES for Small, Passively Heated Buildings

Small passively heated building typically use the soil adjoining the building as a low-temperature seasonal heat store that in the annual cycle reaches a maximum temperature similar to average annual air temperature, with the temperature drawn down for heating in colder months. Such systems are a feature of building design, as some simple but significant differences from 'traditional' buildings are necessary. At a depth of about 20 feet (6.1 m) in the soil, the temperature is naturally stable within a year-round range, if the draw down does not exceed the natural capacity for solar restoration of heat. Such storages operate within a narrow range of storage temperatures over the course of a year, as opposed to the other STES systems described above for which large annual temperature differences are intended.

Two basic passive solar building technologies were developed in the US during the 1970s and 1980s. They utilize direct heat conduction to and from thermally isolated, moisture-protected soil as a seasonal storage medium for space heating, with direct conduction as the heat return method. In one method, "passive annual heat storage" (PAHS), the building's windows and other exterior surfaces capture solar heat which is transferred by conduction through the floors, walls, and sometimes the roof, into adjoining thermally buffered soil.

When the interior spaces are cooler than the storage medium, heat is conducted back to the living space. The other method, "annualized geothermal solar" (AGS) uses a separate solar collector to capture heat. The collected heat is delivered to a storage device (soil, gravel bed or water tank) either passively by the convection of the heat transfer medium (e.g. air or water) or actively by pumping it. This method is usually implemented with a capacity designed for six months of heating.

A number of examples of the use of solar thermal storage from across the world include: Suffolk One a college in East Anglia, England that uses a thermal collector of pipe buried in the bus turning area to collect solar energy that is then stored in 18 100 metres (330 ft) probes for use in the winter heating. Drake Landing Solar Community in Canada uses solar thermal collectors based on the garage roof of 52 homes, is then stored in an array of 35 metres (115 ft) deep probes. The ground can reach temperatures in excess of 70 °C which is then used heat the houses passively. The scheme has been running successfully since 2007. In Brædstrup, Denmark some 8,000 square metres (86,000 sq ft) of solar thermal collectors are used to collect some 4,000,000 kWh/a again stored in an array of 50 metres (160 ft) deep probes.

Liquid Engineering

Architect Matyas Gutai obtained an EU grant to construct a house in Hungary which uses extensive water filled wall panels as heat collectors and reservoirs with underground heat storage water tanks. The design uses microprocessor control.

Small Buildings with Internal STES Water Tanks

A number of homes and small apartment buildings have demonstrated combining a large internal water tank for heat storage with roof-mounted solar-thermal collectors. Storage temperatures of 90 °C (194 °F) are sufficient to supply both domestic hot water and space heating. The first such house was MIT Solar House #1, in 1939. An eight-unit apartment building in Oberburg, Switzerland was built in 1989, with three tanks storing a total of 118 m³ (4,167 cubic feet) that store more heat than the building requires. Since 2011, that design is now being replicated in new buildings.

In Berlin, the "Zero Heating Energy House", was built in 1997 in as part of the IEA Task 13 low energy housing demonstration project. It stores water at temperatures up to 90 °C (194 °F) inside a 20 m³ (706 cubic feet) tank in the basement,.

A similar example was built in Ireland in 2009, as a prototype. The *solar seasonal store* consists of a 23 m³ (812 cu ft) tank, filled with water, which was installed in the ground, heavily insulated all around, to store heat from evacuated solar tubes during the year. The system was installed as an experiment to heat the *world's first standardized pre-fabricated passive house* in Galway, Ireland. The aim was to find out if this heat would be sufficient to eliminate the need for any electricity in the already highly efficient home during the winter months.

Use of STES in Greenhouses

STES is also used extensively for applications as the heating of greenhouses. ATES is the kind of storage commonly in use for this application. In summer, the greenhouse is cooled with ground water, pumped from the "cold well" in the aquifer. The water is heated in the process, and is returned to the "warm well" in the aquifer. When the greenhouse needs heat, such as to extend the growing season, water is withdrawn from the warm well, becomes chilled while serving its heating function, and is returned to the cold well. This is a very efficient system of free cooling, which uses only circulation pumps and no heat pumps.

References

- European Environment Agency (2008). Energy and environment report 2008. EEA Report. No 6/2008. Luxemburg: Office for Official Publications of the European Communities. p. 83. doi:10.2800/10548. ISBN 978-92-9167-980-5. ISSN 1725-9177. Retrieved 2009-03-22.
- "annex 9". National Inventory Report 1990–2006:Greenhouse Gas Sources and Sinks in Canada. Government of Canada. May 2008. ISBN 978-1-100-11176-6. ISSN 1706-3353.
- RETscreen International, ed. (2005). "Ground-Source Heat Pump Project Analysis". Clean Energy Project Analysis: RETscreen Engineering & Cases Textbook. Natural Resources Canada. ISBN 0-662-39150-0. Catalogue no.: M39-110/2005E-PDF. Retrieved 2009-04-20.
- "Environmental Technology Verification Report" (PDF). U.S. Environmental Protection Agency. Archived from the original (PDF) on 2014-04-19. Retrieved December 3, 2015.
- Andersson, O.; Hägg, M. (2008), "Deliverable 10 - Sweden - Preliminary design of a seasonal heat storage for ITT Flygt, Emmaboda, Sweden" (PDF), Deliverable 10 - Sweden - Preliminary design of a seasonal heat storage for ITT Flygt, Emmaboda, Sweden, IGEIA – Integration of geothermal energy into industrial applications, pp. 38–56 and 72–76, retrieved 21 April 2013
- Wong, Bill (June 28, 2011), "Drake Landing Solar Community" (PDF), Drake Landing Solar Community, IDEA/CDEA District Energy/CHP 2011 Conference, Toronto, pp. 1–30, retrieved 21 April 2013
- "Canadian Solar Community Sets New World Record for Energy Efficiency and Innovation" (Press release). Natural Resources Canada. 5 October 2012. Retrieved 21 April 2013. "Drake Landing Solar Community (webpage)". Retrieved 21 April 2013.

10

Toxic Hotspot: An Overview

A toxic hotspot is a location where radiation/discharge from definite sources exposes the local population to high health risks. In urban areas, toxic hotspots can be usually found around polluted emitters such as old factories and waste storage sites. The topic discussed in this chapter are of great importance to broaden the existing knowledge on air pollution.

Toxic hotspots are locations where emissions from specific sources such as water or air pollution may expose local populations to elevated health risks, such as cancer. These emissions contribute to cumulative health risks of emissions from other sources nearby. Urban, highly populated areas around pollutant emitters such as old factories and waste storage sites are often toxic hotspots.

Soil Contamination Hotspots

The 1984 Bhopal disaster in India, the world's worst chemical disaster, is a prime example of a significant toxic hotspot. The toxic gas leaked from the understaffed Union Carbide plant killed up to 20,000 people and left 120,000 others chronically ill. Bhopal continues to face pollution problems from the abandoned factory today.

Air Pollution Hotspots

Air pollution hotspots are areas where air pollution emissions expose individuals to increased negative health effects. Hotspots denote areas in which a population's exposure to pollution and estimated health risks are high. Air pollution hotspots are particularly common in highly populated, urban areas, where there may a combination of stationary sources (e.g. industrial facilities) and mobile sources (e.g. cars and trucks) of pollution. Emissions from these sources can cause respiratory disease, childhood asthma, cancer, and other health problems. Fine particulate matter such as diesel soot, which contributes to more than 3.2 million premature deaths around the world each year, is a significant problem. It is very small and can lodge itself within the lungs and enter the bloodstream. Diesel soot is concentrated in densely populated areas, and one in six people in the U.S. live near a diesel pollution hot spot.

While air pollution hotspots affect a variety of populations, some groups are more likely to be located in hotspots. Previous studies have shown disparities in exposure to pollution by race and/or income (cite one of the early readings from our syllabus, e.g. Mohai & Pellow, or Saha). Hazardous land uses (toxic storage and disposal facilities, manufacturing facilities, major roadways) tend to be located where property values and income levels are low. Low socioeconomic status can be a proxy for other kinds of social vulnerability, including race, a lack of ability to influence regulatory permitting and a lack of ability to move to neighborhoods with less environmental pollution. These communities bear a disproportionate burden of environmental pollution and are more likely to face health risks such as cancer or asthma.

Studies show that patterns in race and income disparities not only indicate a higher exposure to pollution but also higher risk of adverse health outcomes. Communities characterized by low socioeconomic status and racial minorities can be more vulnerable to cumulative adverse health impacts resulting from elevated exposure to pollutants than more privileged communities. Blacks and Latinos generally face more pollution than whites and Asians, and low-income communities bear a higher burden of risk than affluent ones. Racial discrepancies are particularly distinct in suburban areas of the South and metropolitan areas of the West. Residents in public housing, who are generally low-income with poor access to health care and cannot move to healthier neighborhoods, are highly affected by nearby refineries and chemical plants.

Community groups and academic researchers have argued the unequal distribution of pollution on the poor and communities of color is an "environmental justice".

Policy makers and researchers concerned with improving environmental justice for communities situated next to major sources of air pollution have developed a number of regulatory tools to identify air pollution hotspots. The EPA, for example, utilizes their Risk-Screening Environmental Indictors (RSEI) model to identify hotspots from a score of 3 to 15, with higher scores indicating closer proximity to hazards. Individual states have also taken steps to improve identification and surveillance. California's AB 2588 Air Toxics "Hot Spots" Program, enacted in 1987, seeks to collect emission data, determine health risks, and notify local residents of major risks. By identifying hotspots regulators hope these tools will help them reduce pollution and inform nearby populations through the health risk assessments of individual pollutants and facilities that are summed in each zone to develop a total lifetime cancer risk. Air pollution hot spots are also at issue in pollution-trading programs, such as cap-and-trade systems designed to control pollution. These programs can potentially exacerbate effects from air pollution hotspots if the differences in chemical hazards are ignored. These programs also cause pollution to be mitigated towards where credit-buying firms are located. Factories can purchase emissions reduction credits from other firms, which leads to concentrated areas of pollution, since facilities that sell their credits are "exporting" their pollution to firms more likely to buy credits. However, some studies have noted that these claims have not materialized. Evan Ringquist, a professor at Indiana University of Public and Environmental Affairs, states that there is little empirical evidence to suggest the emergence of hotspots.

West Oakland, California

Located in the East San Francisco Bay, the neighborhood of West Oakland is home to mainly low-income, African American and Latino residents who are exposed to a disproportionate amount of airborne toxins, as compared to the rest of the surrounding Alameda County. West Oakland's close proximity to highways and the Port of Oakland leave residents highly exposed to pollutants caused by moving and stationary sources of diesel pollution, thus leaving them at higher risk for health complications such as asthma and even shorter life expectancy than surrounding neighborhoods averages.

High emissions of toxic chemicals and airborne particulate matter in West Oakland that cause health issues are due to diesel fuels used for transportation in the Port of Oakland and surrounding highways. Traffic and transportation related air pollutants include carbon monoxide, nitrogen dioxide, black carbon, and diesel particulate matter. Residents are more exposed to harmful pollutants compared to other areas of the Bay Area and Oakland and therefore more at risk for harmful health effects. Compared to the State of California, West Oakland produces 90 times more diesel emission particulates per square mile per day. These pollutants have detrimental health effects such as asthma and reduced life expectancy while putting children at higher susceptibility for health complications.

Inequitable economic, residential, and environmental conditions in this low-income community of color leave residents of West Oakland with poor and inequitable health outcomes. African-American and Latino children of 10–18 years in West Oakland are more susceptible to onset lung defects such as asthma. According to Alameda County Vital Statistics, an African American child born in West Oakland is expected to live 14 fewer years than a white child born in the more wealthy Oakland Hills. Children 5 and under in west Oakland visit the emergency room for asthma three times more often than children in the county as a whole.

There are multiple efforts and strategies to spur legislation for equitable environmental conditions in low-income communities. There are many environmental justice groups and organizations in the Bay Area that encourage community participation in pursuing environmental justice. For example, data is collected by a Community-based participatory research (CBPR) and collaborated with West Oakland Environmental Indicators Project (WOEIP) in order to find effective and accurate findings to prove injustice and eventually spur reform in environmental policy. These research efforts can be used to document and communicate trends in air quality in West Oakland to policy makers. Effectiveness of efforts by these groups are multiplied by and increasing availability of environmental poverty lawyers who empower legislation in the legal system.

Richmond, California

Richmond, located in the San Francisco Bay Area, is an evolving, multi-cultural community that has transformed itself from an over-polluted industrial town to a pioneer in an environmental justice movement. The city has been host to numerous oil refineries, including the Chevron Corporation refinery, which opened in 1901 under the ownership of Pacific Coast Oil. The Chevron Refinery is a leading source of air quality violations in the state of California. Richmond residents are also exposed to pollution from the Santa Fe train line and the presence of heavy traffic and diesel trucks along the Richmond Parkway. However, residents are most concerned with air pollution health impacts from the Chevron Refinery. In 1999, Richmond measure significantly higher on Air Quality Indices (AQI) (an indicator of how polluted is air is) compared to national level. Air pollution emission from the Chevron refinery includes benzene, ethyl benzene, toluene, xylene, nitrous dioxide, and sulfur dioxide, which are known to cause elevated cancer risks and respiratory illness. Rates of child and adult asthma are especially elevated among Richmond residents.

Richmond residents have struggled to improve local air quality. The city has a significant non-white, low-income population. According to 2010 U.S. Census, of Richmond's 103,701 person populations, "one in six residents lives below the federal poverty level, and more than eight in 10 are people of color. In North Richmond, next to one of the nation's largest refineries, 97 percent of residents are non-white and nearly one in four live in poverty". Low-income communities have differential access to political power, and their collective political voice is often less able to contest decisions impacting industrial operations. The combination of poverty, poor access to clean air, and poor political power can result in inequality in which communities of color bear a disproportionate burden of pollution and, therefore, suffer from greater environmental health risks.

Because Richmond is an air pollution hotspot, Richmond residents have applied different strategies since the 1980s to try to improve local air quality. The first EJ movement in the area started in the late 1980s, when the activist tried to stop construction of a garbage incinerator near North Richmond. Sixteen years later, local citizen utilizes "Bucket Brigades" to document a handful of criteria air pollutants such as sulfur dioxide [SO2], carbon monoxide [CO], nitrous dioxide [NO2], and ozone [O3].This study involves citizens to actively collecting the samples of emissions from Chevron's refineries, especially during accidents, fires, leaks, and explosions.The "sniffers" alert the "samplers" to collect the air samples when they notice a problem.The "samplers" then contact the Coordinator to check the bucket and perform the paperwork before submitting the samples to the Laboratory, in which the results will be reported to CBE, an environmental justice organization. The "Bucket Brigades" did not only raise the awareness local citizens to fight against the air pollution in their area, but also their participation.

As the number of activists and participants grew in numbers, their position in the battle against environmental injustice was further fortified with the election Green-party mayor of Richmond, Gayle McLaughlin, as well as three new council members sympathetic to their cause in 2008. In July 2008, despite the council failure to halt the Chevron's plan to build more refineries in the area due to rising gasoline prices during that time, the council succeeded to acquire $61 million from the oil company for community programs.

Due to great forces from the local communities and fellow EJ activists in Richmond area, Chevron has been making progress to embrace cleaner environment. In 2005, local activists managed to convince Bay Area Air Quality Management District to tighten the air pollution regulations by increasing the frequency of fines of facility incidents. Since then, Chevron has been flaring 10 times less than before. On top of that, Chevron has invested $150 million for building gas turbine in order to reduce air emission, increase energy efficiency, as well as provide most electrical and steam power Chevron requires to operate.

Wilmington, Los Angeles

Bonnoris noted, "The environmental justice movement posits that the distribution of environmental harms and benefits should be fairly apportioned among all communities". As Bonnoris argued, the burden of air pollution is disproportionally distributed among communities based on their racial, social or economic status. Disproportion distribution of air pollution among communities can be a violation of the Equal Protection Clause of the Constitution because it violates equal protection of residents' public health.

Los Angeles is known for the nation's worst air quality and its "sharp inequalities in environmental exposures". Wilmington, Los Angeles is a neighborhood located on the southern part of Los Angeles, California. 54,512 people live in Wilmington, the median household income is $40,627, about 86 percentage of them are Latino and only 5.1% of Wilmington residents 25 or older have a four-year degree.

Wilmington, most of its residents are ethnic minorities, is possible to bear more environmental burden than other communities in Los Angeles because it is located next to several sources of air pollution. For example, Wilmington has "the highest concentration of refineries in the State". Emissions from refineries in Wilmington include carbon dioxide, sulfur dioxide and benzene.

Wilmington has higher concentration of diesel particulate matter due to emissions from diesel trucks from the ports of Los Angeles and Long Beach. The risks associated with diesel are often underestimated since existing epidemiological studies cannot isolate exposure to diesel PM. However, exposure to diesel particulate matter can cause "irritation to the eyes, nose, throat and lungs", asthma, "exhaust immunological effects", and cancer.

Several NGOs have worked to improve the accuracy of Wilmington air quality data and air quality in order to protect approximately 1400 children who live or visit schools or childcare facilities at Wilmington. The environmental group "Coalition For a Safe Environment" installed an air pollution monitoring devices on the residential buildings in Wilmington in order to prove that emissions from local oil refineries and diesel trucks to the ports pollute the air in Wilmington, disproportionately affecting Wilmington residents to suffer from health problems including lung diseases and respiratory diseases.

Groundwater Contamination

The town of Hinkley, California, located in the Mojave Desert, had its groundwater contaminated with hexavalent chromium starting in 1952, resulting in a legal case against Pacific Gas & Electric (PG&E) and a multimillion-dollar settlement in 1996. The legal case was dramatized in the film *Erin Brockovich*, released in 2000.

PG&E operates a compressor station in Hinkley for natural gas transmission pipelines. The natural gas has to be re-compressed approximately every 350 miles (560 km), and the station uses large cooling towers to cool the gas after it has been compressed. Between 1952 and 1966, the water used in these cooling towers contained hexavalent chromium – now recognized as a carcinogen – to prevent rust in the machinery. The water was stored between uses in unlined ponds, which allowed it to percolate into the groundwater. This severely contaminated the groundwater, affecting soil and contaminating water wells near the compressor station, with a plume approximately 2 miles (3.2 km) long and nearly 1 mile (1.6 km) wide.

Radioactive Contamination

Pacific Proving Grounds

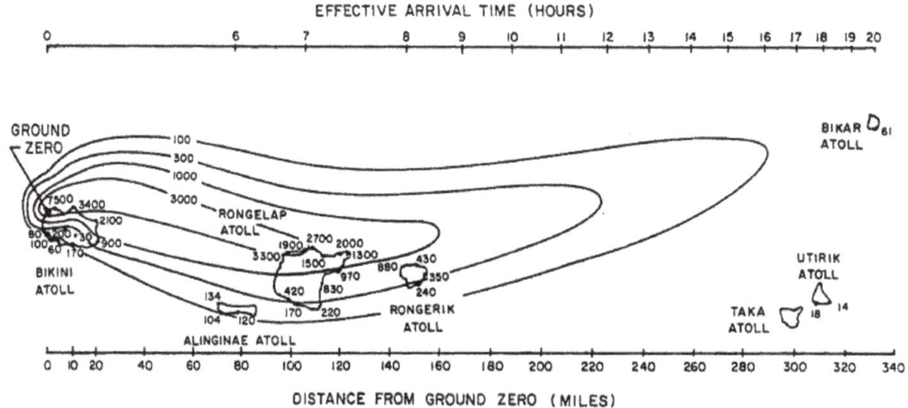

The Castle Bravo test of 1954 spread nuclear fallout across the Marshall Islands, parts of which were still inhabited.

The Pacific Proving Grounds was the name used to describe a number of sites in the Marshall Islands and a few other sites in the Pacific Ocean, used by the United States to conduct nuclear testing at various times between 1946 and 1962. In July 1947, after the first atomic weapons testing at Bikini Atoll, the United States entered into an agreement with the United Nations to govern the Trust Territory of the Pacific Islands as a strategic trusteeship territory. The Trust Territory is composed of 2,000 islands spread over 3,000,000 square miles (7,800,000 km^2) of the North Pacific Ocean. On July 23, 1947, the United States Atomic Energy Commission announced the establishment of the Pacific Proving Grounds.

105 atmospheric (i.e., not underground) nuclear tests were conducted there, many of which were of extremely high yield. While the Marshall Islands testing composed 14% of all U.S. tests, it composed nearly 80% of the total yields of those detonated by the U.S., with an estimated total yield of around 210 megatons, with the largest being the 15 Mt Castle Bravo shot of 1954 which spread considerable nuclear fallout on many of the islands, including several which were inhabited, and some that had not been evacuated.

Many of the islands which were part of the Pacific Proving Grounds continue to be contaminated by nuclear fallout, and many of those who were living on the islands at the time of testing have suffered from an increased incidence of various health problems. Through the Radiation Exposure Compensation Act of 1990, at least $759 million has been paid to Marshall Islanders as compensation for their exposure to U.S. nuclear testing. Following the Castle Bravo accident, $15.3 million was paid to Japan.

Nevada Test Site

Mushroom cloud from the Nevada Test Site seen from downtown Las Vegas.

The Nevada Test Site (NTS), is a United States Department of Energy reservation located in southeastern Nye County, Nevada, about 65 miles (105 km) northwest of the city of Las Vegas. Formerly known as the Nevada Proving Grounds, the site was established on 11 January 1951 for the testing of nuclear devices, covering approximately 1,360 square miles (3,500 km^2) of desert and mountainous terrain. Nuclear testing at the Nevada Test Site began with a 1-kiloton-of-TNT (4.2 TJ) bomb dropped on Frenchman Flat on 27 January 1951. Many of the iconic images of the nuclear era come from the NTS.

During the 1950s, the mushroom clouds from atmospheric tests could be seen for almost 100 mi (160 km). The city of Las Vegas experienced noticeable seismic effects, and the distant mushroom clouds, which could be seen from the downtown hotels, became tourist attractions. St. George, Utah, received the brunt of the fallout of above-ground nuclear testing in the Yucca Flats/Nevada Test Site. Winds routinely carried the fallout of these tests directly through St. George and southern Utah. Marked increases in cancers, such as leukemia, lymphoma, thyroid cancer, breast cancer, melanoma, bone cancer, brain tumors, and gastrointestinal tract cancers, were reported from the mid-1950s through 1980.

From 1986 through 1994, two years after the United States put a hold on full-scale nuclear weapons testing, 536 anti-nuclear protests were held at the Nevada Test Site involving 37,488 participants and 15,740 arrests, according to government records. Those arrested included the astronomer Carl Sagan and the actors Kris Kristofferson, Martin Sheen, and Robert Blake.

The Nevada Test Site contains 28 areas, 1,100 buildings, 400 miles (640 km) of paved roads, 300 miles of unpaved roads, ten heliports, and two airstrips. The most-recent test was a sub-critical test of the properties of plutonium, conducted underground on December 7, 2012.

Semipalatinsk Test Site

The Semipalatinsk Test Site, also known as “The Polygon”, was the primary testing venue for the Soviet Union’s nuclear weapons. It is located on the steppe in northeast Kazakhstan (then the Kazakh SSR), south of the valley of the Irtysh River. The scientific buildings for the test site were located around 150 km west of the town of Semipalatinsk (later renamed Semey), near the border of East Kazakhstan Province and Pavlodar Province with most of the nuclear tests taking place at various sites further to the west and south, some as far as into Karagandy Province.

The Soviet Union conducted 456 nuclear tests at Semipalatinsk from 1949 until 1989 with little regard for their effect on the local people or environment. The full impact of radiation exposure was hidden for many years by Soviet authorities and has only come to light since the test site closed in 1991.

From 1996 to 2012, a secret joint operation of Kazakh, Russian, and American nuclear scientists and engineers secured the waste plutonium in the tunnels of the mountains.

References

- Gerrard, Michael B.; editors, Sheila R. Foster, (2008). Law of environmental justice : theories and procedures to address disproportionate risks (2nd ed.). Chicago, Ill.: American Bar Association, Section of Environment, Energy, and Resources. ISBN 1604420839.
- McDougal, Myres S. and Schlei, Norbert A. “The Hydrogen Bomb Tests in Perspective: Lawful Measures for Security”. In Myres S. McDougal, et al. (1987), Studies in World Public Order, p. 766. New Haven: New Haven Press. ISBN 0-89838-900-3.

- Pettit, David (14 December 2014). "Global Toll of Air Pollution: Over 3 Million Deaths Each Year". Switchboard NRDC.
- Palaniappan, Meena. "Clearing the Air: Reducing Dielsel Pollution in West Oakland" (PDF). Pacific Institute. Retrieved 23 April 2014.
- "List of Air Emissions That Chevron's Richmond Refinery Project Could Increase If Mitigation Is Not Required According To The Revised Draft EIR*" (PDF). Communities for a Better Environment. Retrieved 30 April 2014.
- Brody, Julia Green; Morello-Frosch R; Zota A; Brown P; Pérez C; Rudel RA. (November 2009). "Linking exposure assessment science with policy objectives for environmental justice and breast cancer advocacy: the northern California household exposure study". American Journal of Public Health 99: S600–S609. doi:10.2105/ajph.2008.149088. Retrieved 30 April 2014.
- Morello-Frosch, Rachel A (2002). "Discrimination and the political economy of environmental inequality" (PDF). Environment and Planning C website 20 (4): 477–496. doi:10.1068/c03r. Retrieved 7 May 2014.
- Cheryl Katz; Jane Kay (2012). "'We are Richmond.' A beleaguered community earns multicultural clout.". Environmental Health News. Retrieved 30 April 2014.
- O'rourke, Dara; Gregg P. Macey (2003). "Community environmental policing: Assessing new strategies of public participation in environmental regulation" (PDF). Journal of Policy Analysis and Management 22 (3): 383–414. doi:10.1002/pam.10138. Retrieved 30 April 2014.
- Jane Kay; Cheryl Katz (June 5, 2012). "We are Richmond. A Beleaguered Community Earns Multicultural Clout". Environmental Health News. Retrieved 30 April 2014.
- California Environmental Protection Agency Air Resources Board (November 2013). "Community Air Quality Monitoring: Special Studies Wilmington" (PDF). California Environmental Protection Agency Air Resources Board. Retrieved 24 April 2014.
- National Nuclear Security Administration / Nevada Site Office (January 2011). "Miss Atom Bomb" (PDF). Fact Sheets. Retrieved 2011-12-02.

Permissions

All chapters in this book are published with permission under the Creative Commons Attribution Share Alike License or equivalent. Every chapter published in this book has been scrutinized by our experts. Their significance has been extensively debated. The topics covered herein carry significant information for a comprehensive understanding. They may even be implemented as practical applications or may be referred to as a beginning point for further studies.

We would like to thank the editorial team for lending their expertise to make the book truly unique. They have played a crucial role in the development of this book. Without their invaluable contributions this book wouldn't have been possible. They have made vital efforts to compile up to date information on the varied aspects of this subject to make this book a valuable addition to the collection of many professionals and students.

This book was conceptualized with the vision of imparting up-to-date and integrated information in this field. To ensure the same, a matchless editorial board was set up. Every individual on the board went through rigorous rounds of assessment to prove their worth. After which they invested a large part of their time researching and compiling the most relevant data for our readers.

The editorial board has been involved in producing this book since its inception. They have spent rigorous hours researching and exploring the diverse topics which have resulted in the successful publishing of this book. They have passed on their knowledge of decades through this book. To expedite this challenging task, the publisher supported the team at every step. A small team of assistant editors was also appointed to further simplify the editing procedure and attain best results for the readers.

Apart from the editorial board, the designing team has also invested a significant amount of their time in understanding the subject and creating the most relevant covers. They scrutinized every image to scout for the most suitable representation of the subject and create an appropriate cover for the book.

The publishing team has been an ardent support to the editorial, designing and production team. Their endless efforts to recruit the best for this project, has resulted in the accomplishment of this book. They are a veteran in the field of academics and their pool of knowledge is as vast as their experience in printing. Their expertise and guidance has proved useful at every step. Their uncompromising quality standards have made this book an exceptional effort. Their encouragement from time to time has been an inspiration for everyone.

The publisher and the editorial board hope that this book will prove to be a valuable piece of knowledge for students, practitioners and scholars across the globe.

Index

www.ingramcontent.com/pod-product-compliance
Lightning Source LLC
Chambersburg PA
CBHW061247190326
41458CB00011B/3607

9781635490213